INTERNATIONAL ENERGY AGENCY

GLOBAL ENERGY
THE CHANGING OUTLOOK

INTERNATIONAL ENERGY AGENCY
2, RUE ANDRÉ-PASCAL, 75775 PARIS CEDEX 16, FRANCE

The International Energy Agency (IEA) is an autonomous body which was established in November 1974 within the framework of the Organisation for Economic Co-operation and Development (OECD) to implement an international energy programme.

It carries out a comprehensive programme of energy co-operation among twenty-one* of the OECD's twenty-four Member countries. The basic aims of IEA are:
- i) co-operation among IEA participating countries to reduce excessive dependence on oil through energy conservation, development of alternative energy sources and energy research and development;
- ii) an information system on the international oil market as well as consultation with oil companies;
- iii) co-operation with oil producing and other oil consuming countries with a view to developing a stable international energy trade as well as the rational management and use of world energy resources in the interest of all countries;
- iv) a plan to prepare Participating Countries against the risk of a major disruption of oil supplies and to share available oil in the event of an emergency.

IEA Participating Countries are: Australia, Austria, Belgium, Canada, Denmark, Germany, Greece, Ireland, Italy, Japan, Luxembourg, the Netherlands, New Zealand, Norway, Portugal, Spain, Sweden, Switzerland, Turkey, United Kingdom, United States.

Pursuant to article 1 of the Convention signed in Paris on 14th December 1960, and which came into force on 30th September, 1961, the Organisation for Economic Co-operation and Development (OECD) shall promote policies designed:
- to achieve the highest sustainable economic growth and employment and a rising standard of living in Member countries, while maintaining financial stability, and thus to contribute to the development of the world economy;
- to contribute to sound economic expansion in Member as well as non-member countries in the process of economic development; and
- to contribute to the expansion of world trade on a multilateral, non-discriminatory basis in accordance with international obligations.

The original Member countries of the OECD are Austria, Belgium, Canada, Denmark, France, the Federal Republic of Germany, Greece, Iceland, Ireland, Italy, Luxembourg, the Netherlands, Norway, Portugal, Spain, Sweden, Switzerland, Turkey, the United Kingdom and the United States. The following countries became Members subsequently through accession at the dates indicated hereafter: Japan (28th April 1964), Finland (28th January 1969), Australia (7th June 1971) and New Zealand (29th May 1973).

The Socialist Federal Republic of Yugoslavia takes part in some of the work of the OECD (agreement of 28th October 1961).

© OECD/IEA, 1992
Application for permission to reproduce or translate
all or part of this publication should be made to:
Head of Publications Service, OECD
2, rue André-Pascal, 75775 PARIS CEDEX 16, France.

TABLE OF CONTENTS

FOREWORD 7

EXPLANATORY NOTE 9

A NOTE ON REGIONAL COUNTRY GROUPINGS 11

I. INTRODUCTION
   1. The Fundamental Change in the Global
      Energy Balance 13

II. SUMMARY AND CONCLUSIONS
   1. The Major Changes in Regional and
      Fuel Demand Structures 17
   2. Responding to the Global Energy Security Challenge 21
   3. Overcoming the Barriers to Improved Energy Efficiency
      and Fuel Diversification — the Role of Energy
      Technologies 26
   4. Energy and Environment — the Global Challenge 27
   5. What Role for the IEA? 31

III. RECENT DEVELOPMENTS IN NON-OECD ENERGY SUPPLY
    AND PRODUCTION
   1. Primary Energy Supply 35
   2. Primary Energy Production 44
   3. Electricity Generation 53
   4. Energy Intensity 54

5. The Non-OECD's Exportable Energy Surplus —
the Trade Relationship with OECD Countries 59

IV. THE ENERGY OUTLOOK IN NON-OECD COUNTRIES
1. Introduction 63
2. Total Non-OECD 65
3. Africa 83
4. Asia-Pacific 95
5. China 107
6. Eastern Europe 119
7. Latin America 131
8. Middle East 143
9. USSR 151

V. ENERGY-RELATED ENVIRONMENTAL ISSUES
IN NON-OECD COUNTRIES
1. The International Dimension 163
2. Areas of Environmental Concern 165
3. An Overview of the Major Transboundary and
Global Environmental Problems 167
4. Potential Policy Responses to Environmental Issues 180

STATISTICAL ANNEX 187

GLOSSARY 199

SELECTED BIBLIOGRAPHY 201

TABLES

1. Traditional Fuel Demand in Non-OECD Regions, 1989 44
2. Energy Intensity — Selected Countries, 1973 and 1989 58
3. Non-OECD — Primary Energy Balance, 1989-2005 82
4. Africa — Primary Energy Balance, 1989-2005 94

5.	Asia-Pacific — Primary Energy Balance, 1989-2005	105
6.	China — Primary Energy Balance, 1989-2005	117
7.	Eastern Europe — Primary Energy Balance, 1989-2005	129
8.	Latin America — Primary Energy Balance, 1989-2005	142
9.	Middle East — Primary Energy Balance, 1989-2005	150
10.	USSR — Primary Energy Balance, 1989-2005	162
11.	SO_X and NO_X Emissions in Selected Asian and OECD Countries, 1987	170
12.	Levels and Sources of Sulphur Deposition in Selected European Countries	171
13.	Energy-Related CO_2 Emissions	178

FIGURES

1.	Shares of World TPES	36
2.	Non-OECD TPES by Region	37
3.	Non-OECD TPES by Fuel	39
4.	Shares of World Coal Demand	41
5.	Shares of World Oil Demand	42
6.	Shares of World Gas Demand	43
7.	Shares of World Primary Energy Production	45
8.	Non-OECD Energy Production by Region	46
9.	Non-OECD Energy Production by Fuel	47
10.	Shares of World Coal Production	48
11.	Shares of World Oil Production	49
12.	Shares of World Gas Production	50
13.	Shares of World Nuclear Power Production	51
14.	Shares of World Hydropower Production	52
15.	Inputs to Electricity Generation	54
16.	Energy Intensity	55
17.	Net Energy Trade	60
18.	Shares of World TPES, 1989-2005	68
19.	Non-OECD TPES by Fuel, 1989-2005	69
20.	Non-OECD TPES by Region, 1989-2005	70
21.	Shares of World Energy Production, 1989-2005	71
22.	Non-OECD Energy Production by Fuel, 1989-2005	72
23.	Africa — TPES, 1989-2005	84
24.	Africa — Energy Production, 1989-2005	85
25.	Asia-Pacific — TPES, 1989-2005	97

26.	Asia-Pacific — Energy Production, 1989-2005	98
27.	China — TPES, 1989-2005	109
28.	China — Energy Production, 1989-2005	110
29.	Eastern Europe — TPES, 1989-2005	122
30.	Eastern Europe — Energy Production, 1989-2005	123
31.	Latin America — TPES, 1989-2005	132
32.	Latin America — Energy Production, 1989-2005	133
33.	Middle East — TPES, 1989-2005	145
34.	Middle East — Energy Production, 1989-2005	146
35.	USSR — TPES, 1989-2005	153
36.	USSR — Energy Production, 1989-2005	154

FOREWORD

The industrialised nations traditionally have purchased and consumed the vast majority of all forms of energy entering international trade. Now, however, the continued rapid growth in consumption by many non-OECD countries, and these countries' increasing prominence in world markets, have become leading factors destined to have a fundamental influence upon the global energy equation during the next decade and a half. Already, about half of the world's energy consumption occurs outside the OECD area in regions where vigorous economic growth is demanding significantly greater energy inputs.

Recognition of an impending shift in the proportions of overall supplies to be shared between "developed" and "developing" countries will thus be essential in gauging the challenge to be faced by both producers and consumers of energy in coming years. In continuing their pursuit of energy security and environmental protection, OECD countries will find these inter-related objectives can no longer be treated in isolation from parallel security and environmental claims being registered among swiftly evolving economies in the non-OECD world. The world's major energy exporting countries will also need to factor these new realities into their policies, as they gauge not only the prospective volumes, but also the prospective sources of future demand.

This study examines developments in non-OECD energy supply and demand over the period from the early 1970s and it considers likely developments to the year 2005. In so doing, it also examines the relationship between OECD and non-OECD energy exporting and importing countries in the areas where better communication in dealing with energy problems will be important. This is especially the case in responding to mutually recognised environmental problems of an international or global nature, along with relevant questions regarding energy efficiency and technology transfer.

This work is published under my responsibility as Executive Director of the International Energy Agency. It has been prepared by the IEA Secretariat. The study contemplates the future; it does not necessarily reflect the views of IEA Member governments.

Helga Steeg
Executive Director

EXPLANATORY NOTE

The numerical projections included in this report are derived from the IEA's medium term global energy model which has been in use as an analytical tool by the IEA since 1986. A detailed description of the model and its methodology is beyond the scope of this note and further details may be obtained from the Secretariat. However, it must be emphasised that in addition to the usual warnings accompanying any projections, such as the dependence on some key assumptions, the model's projections included in this report are subject to further qualifications. Hence, the projections should be used only as a reference point for the discussion of trends. There are many reasons for this: first, since the main objective of the report is to discuss the underlying energy trends of non-OECD regions, the projections are based on just one of the IEA's scenarios of energy prices and economic growth. Thus, no sensitivity analysis is presented in this report on alternative assumptions on economic growth, perhaps the most important determinant of energy demand in many developing countries. Secondly, the regional projections are based on information as it was available at the end of 1990 and are compatible with the IEA's 1991 Energy Outlook. A revised version of the Energy Outlook will be available in the first half of 1992 and this is likely to include some revisions to the non-OECD projections.

Finally, there are many special factors that make the modelling of energy trends in non-OECD regions especially difficult and the derived projections more uncertain than is usually the case. These factors include poor quality data for some countries, and frequent structural, political and institutional changes which can greatly affect energy developments. The recent changes in the Soviet Union and Eastern Europe provide a timely reminder of the importance of these factors and the large amount of uncertainty that should be attached to any single numerical projection.

The above caveats should not detract from the key conclusions of this report which do not depend on the exact value of any single number and which are unlikely to be substantially altered by subsequent revisions.

A NOTE ON REGIONAL COUNTRY GROUPINGS

For the purposes of this report, non-OECD countries have been aggregated into five regional groups plus the USSR and China. The five groups are as follows:

1. **Africa** — Algeria, Angola, Benin, Cameroon, Congo, Egypt, Ethiopia, Gabon, Ghana, Ivory Coast, Kenya, Libya, Morocco, Mozambique, Nigeria, Senegal, South Africa, Sudan, Tanzania, Tunisia, Zaire, Zambia, Zimbabwe, other Africa

2. **Asia-Pacific** — Bangladesh, Brunei, Hong Kong, India, Indonesia, Malaysia, Myanmar, Nepal, North Korea, Pakistan, Philippines, Singapore, South Korea, Sri Lanka, Taiwan, Thailand, Vietnam, other Asia-Pacific

3. **Eastern Europe** — Albania, Bulgaria, Czechoslovakia, eastern Germany[1], Hungary, Poland, Romania, Yugoslavia

4. **Latin America** — Argentina, Bolivia, Brazil, Chile, Colombia, Cuba, Ecuador, Guatemala, Jamaica, Mexico, Netherlands Antilles, Panama, Paraguay, Peru, Trinidad, Uruguay, Venezuela, other Latin America

---

1. For technical reasons, the former German Democratic Republic has been retained in eastern Europe and excluded from the OECD.

5. **Middle East**   Bahrain, Iran, Iraq, Israel, Jordan, Kuwait, Lebanon, Neutral Zone, Oman, Qatar, Saudi Arabia, United Arab Emirates, Yemen

The term "developing countries", when used in the report, refers to all non-OECD countries with the exception of the USSR and eastern Europe.

I. INTRODUCTION

THE FUNDAMENTAL CHANGE IN THE GLOBAL ENERGY BALANCE

As indicated in the body of this report, energy demand in non-OECD countries was almost equal to that of the OECD group in absolute terms in 1989. The substantially higher growth rates in the non-OECD region will ensure that, by the mid-1990s, these countries will account for more than half of world energy demand. Together with the increasing dominance of non-member countries in world energy production, this fundamental alteration in the global energy balance is essential to an understanding of likely energy developments in the coming years. It will be a necessary consideration of energy policy makers in their efforts to assure global economic growth while maintaining secure energy supplies and protecting the environment. The implications of this change will be complicated, however, by the increased uncertainties and interdependence in international political and economic relations which make the policy responses of industrialised countries, in particular, more difficult.

The fundamental changes in the world energy balance will have potentially significant implications for the way in which OECD countries deal with their energy requirements. It is unlikely, however, that the major energy challenge which these countries have faced over the preceding years, i.e. the assurance of secure energy supplies, will change. As a net importer of energy the OECD has long been faced with the security issue and has sought to deal with it in a number of ways. In particular, it has established emergency measures to deal with potential oil supply disruptions, has encouraged the diversification of fuel sources, the development of improved energy conversion and end-use technologies and has fostered improvements in energy efficiency. These will continue to be important planks of OECD energy policy and probable developments in the energy outlook are likely to intensify efforts towards their achievement. Of particular significance in this regard will be the continued importance of oil as a primary energy source, the increased concentration of available and "economic" oil supply capacity in the Middle East Gulf region and the likely ongoing fragility of political and economic relations in this part of the world.

Apart from underlying energy developments, recent international events and trends will also shape the manner in which the energy security situation is perceived by OECD countries and the way in which they will respond to it. These factors include the political and economic reconstruction of the Soviet Union and eastern Europe, the reshaping of the Middle East in the aftermath of the Gulf war, and the adjustments of economic growth rates now underway in what has traditionally been known as the industrialised world. All these factors could provide both the impetus and the opportunity for OECD countries to enhance their relations with non-members. A disturbing development, however, has been the increasing respectability given to the notion that energy security and market "stability" ought to be sought in a regional rather than a global context. "Regionalisation" initiatives which seek to restrict market access, either directly or indirectly, to a specific group of countries, however defined, will inevitably lead to trade distortions and associated losses in economic efficiency. These losses may not be identified easily in a climate of low oil prices. With the expected pressure on energy supplies in the future, however, they will become increasingly evident.

In addition to the "traditional" concern of energy security raised above, however, are two issues which will become increasingly important, both of their own accord and because of their potential to complicate the security question. The first of these is the possible impact of growing environmental concerns on energy production and use. Energy-related environmental issues range from localised problems of urban air pollution, through transboundary issues such as acid rain, to genuinely global questions including ozone depletion and climate change. The latter issue in particular is being debated actively at the international level but the outcome of these discussions is not yet clear, adding a further element of uncertainty to the global energy outlook.

The environmental issue also serves to highlight the second of these "new dimensions" and that is the increasing commonality of interest between energy producers and consumers and between competing groups of consumers. The increased competition for available energy supplies, for example, is likely to demand greater co-operation in areas such as the exploration and development of energy reserves. The climate change implications of greenhouse gas emissions illustrate clearly the futility of any one country or even group of countries acting individually — the solution to the problem will require a degree of international co-operation not yet witnessed in the energy or environmental arena.

The main implications of these changes for the OECD are that, in seeking sustainable energy strategies for the coming decade, it will not be possible to remain isolated from developments in non-member countries. For OECD countries this change will demand a possibly significant shift in policy-making perspective, away from one where the primary concern is energy policies in individual member countries towards a global view which recognises that developments in non-member countries will have an equally important impact on the global energy balance and, hence, the energy situation in OECD countries. The outcome of this shift to a global perspective will include an increased awareness that member country objectives relating to energy security and the environment will be enhanced by the restructuring and improved efficiency of non-member energy systems. This is likely to lead to increased linkages between the two regions, particularly in the areas of investment, technology sharing or more direct co-operation agreements, especially in relation to environmental problems.

II. SUMMARY AND CONCLUSIONS

THE MAJOR CHANGES IN REGIONAL AND FUEL DEMAND STRUCTURES

Before analysing how OECD countries might best respond to the new challenges of energy security, environmental protection and interdependence with non-member countries, it is useful to summarise the major changes in regional and fuel demand structures which are expected over the period to 2005. While oil's share of both OECD and non-OECD energy demand will continue the decline evident in recent years, it is expected to remain the dominant fuel, accounting for 35 per cent of global energy demand. Its weight in the various regions is expected to vary from 26 per cent in the USSR and eastern Europe to 36 per cent in the developing countries and 40 per cent in the OECD. The impact of declining oil production levels in the USSR in the medium term and rapidly rising demand in the Asia-Pacific region will be to increase the global competition for available oil supplies. Africa and Latin America may become alternative competitive suppliers to the Middle East but this will depend on the future path of oil prices. It is inevitable that the Middle East region will continue to dominate global oil production and that OECD countries will rely on this region for up to 70 per cent of their oil imports.

Coal is expected to remain an important source of energy in the non-OECD area, particularly in the USSR, eastern Europe, China and parts of the Asia-Pacific region. Growth in consumption is expected to be most vigorous, however, in the two latter regions where the power generation sector will remain the dominant consuming sector. The non-OECD is also likely to increase its share of global coal production to 63 per cent by 2005. The major constraint on the increased use of coal at the global level will be concerns regarding the environment. These will initially be felt to a greater degree in OECD countries but environmental constraints will also operate to an extent in certain non-member areas. The extent to which coal consumption in non-member countries is affected by environmental considerations will probably depend upon international agreements and their success in including non-OECD countries in any binding arrangements.

Natural gas is expected to be the fastest growing component of non-OECD energy demand in the period to 2005 and will represent about 25 per cent of total primary energy supply (TPES) in that year. The highest rate of growth in consumption is expected to occur in developing countries, especially the Middle East, where gas is being substituted for oil in power generation and industry in order to free oil for export. In the USSR and eastern Europe, demand growth will also be strong although its expansion will be constrained by lack of investment in gas consuming equipment. Despite this expansion in gas consumption, there is likely to remain considerable potential for further development. Natural gas resources have been discovered in over fifty developing countries but in the majority of these its exploitation has been restricted by a variety of institutional constraints relating to legislation, regulation, fiscal regimes and financial flows. With the likelihood that gas will become increasingly attractive on both economic and environmental grounds, however, it is possible that these constraints will be overcome. The shares of nuclear and hydropower in total energy demand are expected to remain fairly constant in non-OECD countries and to account for a relatively small proportion of total energy consumption.

At the regional level, although the USSR and China will remain by far the largest consumers of energy in the non-OECD world, the major growth in energy demand is expected to occur in the Middle East and the Asia-Pacific region. In the Middle East this will be accompanied by substantial increases in oil and gas production, more than sufficient to fuel the rapid expansion of a domestic energy-based industrial sector. As a result, the Middle East's potential as a supplier of energy to the OECD is likely to increase. In the case of oil, the Middle East will play an increasingly significant role as a supplier of incremental world demand. By 2005, about 43 per cent of world oil production will be sourced from this region and its exportable surplus will have more than doubled. At the same time OECD countries are expected to source a greater proportion of their oil imports from the region, increasing the economic linkages between the two areas.

In the Asia-Pacific region, energy demand will be fuelled by rapid rates of growth in both economic output and the population level. Coal, and to a decreasing extent, oil will remain the principal elements of the fuel structure and gas will become more important over time. As a result of its limited indigenous oil and coal reserves, the region is expected to become increasingly dependent on imports of these fuels. Net imports of oil, for example, are expected to account for over half of oil consumption by 2005 and those of coal for about 30 per cent of consumption. As a consequence, the Asia-Pacific countries will represent the fastest growing source of demand for available global supplies of these fuels, especially oil. Hence, the energy policies and practices pursued in the region, particularly in relation to energy efficiency and fuel diversification, and their impact on energy demand, will have some bearing on the OECD's ability to gain access to international energy resources.

In Africa, a small number of countries account for the major share of commercial energy demand and production. In the remaining countries, traditional (or non-commercial) forms of energy represent a substantial proportion of energy supply. The shift from traditional to commercial fuel sources is likely to continue in line with the economic development process which, combined with high rates of population growth, will lead to sustained increases in energy demand. Despite these demand pressures, it is expected that the major African countries will generate an increasing exportable oil and gas surplus. With the likely re-integration of South Africa into the international trading environment it is also possible that the region will become an increasingly important supplier of coal to the international market.

The Latin American region is also expected to increase its exportable energy surplus, mainly of oil but with potential also in the coal market. Because of its geographic proximity, the implications of this development for OECD countries are likely to be strongest for the North American economies and it is possible that Venezuela, in particular, will form a viable source of supply to this region.

In the USSR, the major political and economic changes taking place create an enormous degree of uncertainty about the future. At least the next several years will be a period of intense transition with the impact on the energy sector being determined largely by the success of the economic reform process. However, although frequently hampered by outdated equipment and practices and shortages of investment funds, the USSR remains the world's largest producer of energy. Vast reserves of coal, oil and gas remain untapped. Assuming the ultimate success of the reform programme it is possible that the USSR will become a larger player in international energy markets, especially as it shifts its currently substantial trade with the eastern European region towards alternative, hard-currency markets.

Eastern Europe's dependence on imports of energy is expected to increase in the medium term as demand growth will consistently outpace increases in indigenous production. As a result of widespread economic and structural reforms, the eastern European countries are expected to play a greater role in international energy markets and, in a global sense, will represent an additional element of competition for available resources. This effect will be reinforced by the pricing of Soviet exports to the region at international prices. Coal is expected to remain the major element in the eastern European energy balance and, as a consequence, environmental problems are likely to be significant.

Projected economic and population growth in China and its concomitant energy requirements will ensure that the country remains a dominant factor in global energy consumption. China's share of world TPES is expected to reach 10 per cent by 2005 — more than the entire Asia-Pacific region. Although environmental issues are becoming an increasingly important concern for policy makers, coal is expected to continue to dominate the fuel structure and is likely to account for about three-quarters of TPES over the outlook period. While China's oil and coal exports have been an important source of hard currency earnings, the country's exportable energy surplus is expected to be squeezed by increased domestic demand.

RESPONDING TO THE GLOBAL ENERGY SECURITY CHALLENGE

One of the major conclusions which can be drawn about the future structure of global energy consumption is undoubtedly that there will be increasing pressure on available energy resources. This increasing pressure or competition for energy supplies will effect all regions, whether net exporters or importers of energy. The implications for energy security will also be widespread. While OECD countries have already developed policies and mechanisms to deal with the security issue, in the majority of non-OECD countries the problem has yet to be dealt with comprehensively. This is all the more important as non-OECD countries are expected to account for by far the larger proportion of incremental energy demand in the coming years. Hence, a decline in the rate of increase in energy demand in non-OECD countries would do more to enhance global energy security than a similar decline in OECD countries. However, the demands of economic and social progress in the non-OECD world are likely to leave few alternatives to increased energy consumption. Indeed it would be unjustifiable, even if possible, for the developed world to suggest that the lower-income countries forego opportunities for growth through the expansion of energy use.

Improved Energy Efficiency as an Alternative Supply of Energy

An area which offers major opportunities to achieve development objectives while containing growth in energy consumption is through improving the efficiency of energy use. An improvement in energy efficiency occurs when less energy is required to produce a given output. To this extent an improvement in efficiency can be considered as an alternative "supply" of energy which can be "produced" by consumers. It has the additional advantage of being environmentally benign. This view of efficiency as an alternative energy supply is one which has underpinned OECD-country policy-making for many years. The potential impact of efficiency improvements on global energy demand levels, however, is much greater in non-member countries. This is, first, because of their greater contribution to incremental energy demand and, secondly, because many of the opportunities for efficiency gains in OECD countries have already been realised. The achievement of global energy security and environmental objectives will be enhanced, therefore, if efforts to improve efficiency are increased in the non-member world.

Although limited, the rate of change in energy intensity is the best indicator of overall progress towards improved energy efficiency. The impact on energy demand of structural change in economic output will also be captured by this measure, however, and may act to reduce or even outweigh the demand effects of efficiency improvements. The experience of OECD countries in achieving efficiency improvements can have an important demonstration effect in relation to non-members. Since the oil price increases of 1973-74 and 1979-80, IEA countries have intensified their efforts to increase the efficiency with which they consume primary energy resources. For the region as a whole, for example, energy intensity declined by 20 per cent between 1973 and 1985. As a consequence, primary energy demand was about 800 Mtoe less than if energy intensity had remained unchanged. This experience highlights well the advantages of an energy conservation strategy which the IEA has previously identified[1], namely that:

— energy conservation will extend the availability of energy resources that are depletable;

— with a return to tightening energy markets, energy conservation will delay and lessen its impact;

— energy conservation reduces the environmental consequences of energy production and use in a way which is consistent with energy policy objectives;

— investment in energy conservation at the margin provides a better return than investment in energy supply;

— investment in energy conservation can often be undertaken in small increments and is therefore flexible at a time when the energy outlook is uncertain.

---

1. International Energy Agency [IEA] (1987), *Energy Conservation in IEA Countries.* OECD, Paris.

Given that the potential increases in energy demand in the non-member world are greater than those in OECD countries in both absolute and proportional terms, the above advantages of an energy conservation strategy are also likely to be larger. This will be true at both the global level through the impacts of increased efficiency on energy security, the availability of resources and the environment, as well as for individual countries. In developing countries in particular, the ability of energy conservation projects to slow the increase in energy consumption can have substantial impacts on the demand for scarce capital resources. This is especially so in the power generation sector where capital requirements are large and can divert funds away from the achievement of other pressing social and economic objectives.

The experience of non-member countries in the field of energy conservation has been more limited than that of members. This is not to suggest that conservation efforts have not been made or that efficiency gains have not been achieved in individual countries. However, it is widely accepted that substantial potential exists in many countries to increase the efficiency with which energy is consumed, especially in the electricity generation sector. ESMAP[1], for example, has estimated that at current energy prices and with the present state of technology, about 20 per cent of commercial energy consumed by the existing capital stock in developing countries could be saved. Again, it is worth noting that efficiency improvements are just one of the set of factors which will affect the outlook for future energy demand. Gains here may be offset by a range of other factors, including changes in the structure of economic output.

There are various reasons why efforts to improve energy efficiency have been less widespread and less effective in non-member countries than in those of the OECD. Barriers to efficiency can be classified into two broad groups — those relating to energy pricing policies; and other structural impediments. Perhaps the most significant way in which government policies can impede improvements in energy efficiency is through uneconomic energy pricing policies. The frequent underpricing of energy, especially electricity, in non-member countries as well as a range of other economic distortions can bias decision-making away from

---

1. ESMAP (1989), *Energy Efficiency Strategy for Developing Countries: The Role of ESMAP*. Paris.

investments in energy efficiency improvements towards investment in new energy supply capacity. It also distorts individual consumer decisions by encouraging investment in, for example, cheaper but less efficient appliances.

Of the other structural factors which can impede efficiency improvements, one which is particularly significant in many countries is a lack of capital for investment in energy saving equipment and processes. This capital deficiency is often a result, of course, of inefficient pricing policies which do not permit the generation of internal investment funds. It is also frequently due to inadequate national wealth and the inability to generate sufficient hard currency income to permit the servicing of imported capital. The large overhanging debt burdens of many developing countries will be an important constraint in this regard for the foreseeable future. Other structural barriers to efficiency include a lack of information about available technologies and the success of conservation programmes in other countries. In some cases, there is also an information deficiency regarding detailed energy consumption patterns and, hence, the potential for efficiency gains. In the lower-income developing countries the skills necessary to implement effective energy efficiency programmes are also often not available. This can occur at the investment planning stage, where decisions to invest in new equipment or in efficiency improvements are critical, or at the technical implementation stage of efficiency projects. Barriers to international trade can also be important impediments to efficiency improvements in developing countries as the necessary technology is frequently not available from domestic sources and is often subject to high tariffs and other restrictive measures. Organisational structures can also be problematic. OECD experience indicates that the successful implementation of efficiency measures is enhanced where authority for energy conservation is visible and has adequate resources to pursue its objectives. This is frequently not the case in non-member countries, particularly where responsibility for energy policy is disaggregated between separate fuel-specific agencies, with no over-riding conservation authority.

Although the impediments to improved energy efficiency are pervasive in many countries, they are not insurmountable, as the experience of a number of countries demonstrates. As discussed earlier, the potential gains from efficiency improvements in non-member countries will be appropriated not only by the countries themselves but also by OECD countries in the form of enhanced energy security. As a result, it is important to consider the role that OECD countries might play in helping to realise these opportunities. Two important areas of assistance will be the dissemination of information and the provision of direct financial resources to assist in the implementation of efficiency programmes. As discussed below, information on available technologies will be particularly important, as will that concerning the economic and technical opportunities for efficiency improvements and the outcomes of efficiency programmes in other countries. Financial assistance will often be a prerequisite to the implementation of programmes, particularly in relation to the development and transfer of appropriate technologies.

Fuel Diversification Strategies

The second major mechanism by which global energy security might be enhanced is the continued diversification of fuel sources away from oil. In OECD countries, the share of oil in the fuel structure has fallen considerably since the first price shock of 1973-74 but the trend has been less pronounced in the non-OECD region. Oil still represents about 35 per cent of non-member primary energy consumption and is likely to be about 32 per cent in 2005. Given the current level of available energy technology and the continuation of current environmental policies, it is unlikely that fuel diversification outside these limits will occur in non-OECD countries. Within the OECD, however, governments are giving increasing attention to renewable energy technologies in recognition of the sizeable potential contribution they can make in the longer term to furthering energy security and environmental protection. In the case of non-OECD countries, the increased utilisation of renewables could play a substantial role in achieving the same objectives. A major impediment to the expansion of renewables in these countries is the lack of accessibility to appropriate technologies at economic prices.

It is worth noting the particular role of nuclear power in efforts to diversify the fuel structure. Since 1973, nuclear power has made a significant contribution to meeting rising electricity demand in OECD countries and in reducing dependence on oil for power generation. Since the late 1970s, however, fewer orders for new plants have been placed in OECD countries, stemming in part from concern about the consequences of nuclear accidents, the lack of adequate methods of disposal of nuclear waste and the costs, including decommissioning costs, of nuclear power plants. Similar trends have become evident in non-OECD countries as well, including some countries where the growth in nuclear output has been substantial. Continuing constraints on the development of nuclear power, especially if compounded by limitations on the use of coal and hydropower for environmental reasons, could seriously hinder efforts to diversify fuel sources away from oil.

Despite the obvious advantages of diversifying the fuel mix, it would be a mistake to consider fuel diversification strategies from this point of view alone. Energy security will be enhanced equally by strategies to diversify the sources of supply of any particular fuel but particularly of oil. The energy security implications of heavy dependence on oil, for example, can be reduced by developing a range of alternative oil supply sources. This approach to energy security raises various questions for both OECD and non-member countries, including the appropriate level of energy exploration efforts, the role of foreign investment in facilitating this and the existing institutional impediments, including price factors, to foreign participation in many resource-rich but capital-poor non-member countries.

OVERCOMING THE BARRIERS TO IMPROVED ENERGY EFFICIENCY AND FUEL DIVERSIFICATION — THE ROLE OF ENERGY TECHNOLOGIES

Much of the work on improving energy technologies, particularly in the area of basic research, is undertaken in the industrialised countries of the OECD. The near absence of research and development (R&D) activities in developing countries is usually the result of a lack of financial resources and, often, the necessary technical skills. As a consequence, non-OECD countries are highly dependent on the development and dissemination of technology by member countries. This raises a number of questions regarding the adaptability of industrialised country technology for non-member country purposes, as well as the accessibility of the latter to new energy technologies. In the first case, it is often found that the application of new technologies in developing countries can only be achieved after a considerable amount of adaptation to local conditions. Equipment in developing countries, for example, often needs to be capable of operating with minimal maintenance and of withstanding fluctuating voltages in power grids. Many technologies also need to be demonstrated in order to establish operating parameters and costs. As a result of deficiencies in these factors, many proven energy technologies are not as widely used as they might be and, hence, the potential gains from technology improvements are not being realised. While a variety of new technologies might be available to developing countries, their "accessibility" is often limited. A major factor constraining accessibility is the lack of information transfer mechanisms such as training programmes. The establishment of such programmes is an important means of developing indigenous capacity to undertake technology assessments and demonstrations and can have significant long-term benefits. A second component of information transfer is the ease with which information on energy technologies can be identified and disseminated. Thirdly, in many developing countries, the institutions necessary for R&D to make a contribution are often ineffective or even missing. This is particularly the case where there is no efficiently functioning system of patents and intellectual property rights. Many industrialised-country suppliers of energy technologies have been reluctant to provide technology to countries without the protection provided by a patent system. Until some of these barriers are overcome, the potential contribution of new technologies to energy efficiency and fuel diversification will remain substantially under-realised.

ENERGY AND ENVIRONMENT
THE GLOBAL CHALLENGE

While the primary energy policy concern of OECD countries over a number of years has been the security of energy supplies, attention over the 1980s has focused increasingly on environmental objectives.

At their meeting in June 1991, for example, IEA Ministers reaffirmed their governments' commitment to develop integrated policies with the objectives of energy security, environmental protection and economic growth. Part of the background to the concern with environmental issues has been the recent international discussion on the links between economic activity, economic growth and the environment, as well as the notion of sustainable economic development. The essence of this idea is that development should be such as to meet the needs of the present generation without compromising the ability of future generations to meet their needs. It also embodies the view that the natural environment, i.e. air, water, land and the eco-system, is a limited resource which can be depleted to the detriment of future generations. For the world energy industry these ideas have been understood to mean that energy should be available at prices moderate enough to ensure that economic growth is accomplished with acceptable environmental impacts. An outcome of this debate is that environmental factors have become important determinants of OECD energy policies and will continue to influence the development of sustainable energy strategies.

OECD countries have already tackled a range of environmental consequences of energy use with considerable success. These include problems of urban air quality arising from emissions of particulates, sulphur dioxide (SO_2) and nitrous oxides (NO_X), water and soil contamination and acid deposition. The current concérns in the environmental area have shifted to include issues of a global nature, particularly stratospheric ozone depletion and the implications of emissions of greenhouse gases for global climate change. The environmental priorities of non-member countries are likely to be different from those of OECD countries to the extent that these countries are now facing the environmental problems that began to be addressed in the industrialised world twenty years ago. Because of its potentially direct impact on energy consumption, however, this study has largely restricted its attention to the question of greenhouse gas emissions and, in particular, to emissions of carbon dioxide (CO_2). While there are obviously many implications of the discussions on greenhouse gas emissions, two of particular relevance are (i) the extent to which policy responses to global climate change will affect levels and patterns of energy consumption and the impact of this on global energy security; and (ii) the implications for relationships between developed and developing countries.

International responses to global climate change are currently the subject of debate within the context of the Intergovernmental Negotiating Committee on the Framework Convention for Climate Change (INC). The outcome of these discussions is far from clear but is likely to include some commitment, at least by developed countries, to reduce emissions of greenhouse gases from fossil fuel combustion. This might be achieved by international agreement and regulation or it is possible that such responses might be implemented through a system of taxes on, for example, the carbon content of fuels. Regardless of the means of achieving the desired outcome, decisions reached will probably have an impact on both the level and fuel structure of energy demand. Any measures which alter the relative attractiveness (whether by price signals through taxes or by administrative regulation) of the different fuels will lead to a shift in energy demand away from carbon-intensive fuels (especially coal), towards less carbon-intensive alternatives. One important consequence in the short-term is likely to be increased consumption of oil. This will occur because there will be substantial time lags between the shift in relative prices and the construction of alternative energy supply capacity. In the medium term, natural gas is likely to play a larger role and the attractiveness of nuclear, hydropower and renewables will increase. This factor will inevitably complicate the energy security issue, however, by increasing, at least in the short term, the dependence of oil deficient economies on the oil exporting regions, principally the Middle East.

These impacts will be modified, of course, if measures taken are not restricted to the carbon content of fuels but are extended to all or a greater range of greenhouse gases, or else are related to broader aggregates such as total energy consumption. As discussed elsewhere in the report, there are major conceptual and methodological problems in measuring the emissions of most non-CO_2 greenhouse gases, and the impacts of a more broadly based tax or regulatory structure on interfuel substitution and, hence, energy security issues are difficult to predict. Even further complications would be added if the environmental debate were not restricted to global climate change issues, but included all environmental effects associated with energy consumption. For example, the environmental advantages of nuclear and hydropower over fossil fuels may not be so explicit if broader environmental issues are included in the analysis. These include safety risks, waste disposal and decommissioning problems in the case of nuclear and questions such as the loss of forests, disruption of biosystems, soil erosion and the silting of rivers and delta lands in the case of hydropower.

The importance of the second point raised above derives from the fact that the resolution of the greenhouse gas emissions problem will require concerted action on an international, if not a global, scale. Action by OECD countries alone, although not to be discouraged if only for its demonstration effects, will be less effective than co-ordinated policy measures by all countries responsible for the major share of emissions. This would require the co-operation of, at least, the USSR, China and India, who together account for about 30 per cent of world CO_2 emissions. The major challenge for participants in the INC process, therefore, is to devise strategies which will be acceptable to all relevant countries. Given the different environmental and economic priorities of those involved, this will demand a degree of flexibility on behalf of all countries. OECD countries, for example, will be required to recognise the particular difficulties that a reduction, or even a slowing in the rate of increase, of CO_2 emissions will impose on developing countries where economic development is likely to be closely tied to increases in energy consumption. Solutions will probably require assistance from the developed to some of the capital-constrained developing countries to aid in the achievement of global objectives. It is certainly worth remembering in this context that the marginal return on an OECD dollar invested in emission control is likely to be higher in non-OECD countries than in the OECD. Financial flows can, therefore, be rationalised on an economic basis and should not be viewed simply as assistance.

The role of improvements in energy efficiency should also not be underestimated in the achievement of global environmental objectives. The driving forces for efficiency improvements to date have been economics, energy security considerations and the development of new technologies rather than environmental concerns. More recently, however, much greater interest has been generated in energy efficiency as a vehicle for achieving environmental objectives. In the case of greenhouse gas emissions, improved efficiency and interfuel substitution are, given current technologies, the only available, cost-effective mechanisms for attaining these goals.

WHAT ROLE FOR THE IEA?

Analysis presented later in this report makes it abundantly clear that, in seeking to improve their energy security, economic growth prospects and environmental protection, OECD countries will be best served by recognising the mutuality of their interests with those of non-members. The increasing interdependencies in the global energy system will ensure that OECD interests will be met by assisting non-members to develop efficient and environmentally sound energy delivery and end-use systems. In recognising these areas of self-interest, assistance is likely to be forthcoming at a number of levels and will inevitably involve a range of multi-lateral and bi-lateral arrangements. Individual IEA member countries will undoubtedly participate in these initiatives. In seeking an "institutional" role for the IEA, however, the Agency will be limited by the fact that it is neither an aid organisation nor a financial institution and that its resource base is small compared with other international bodies.

Until recently, the IEA has concentrated its efforts on matters of direct policy concern to its member countries. In doing so it has demonstrated two valuable characteristics that will continue to serve it well in meeting new challenges. The first of these is the *cohesiveness* it has preserved in its policy responses, often in the face of divergent national interests. This cohesiveness was most recently displayed throughout the Gulf crisis where co-ordinated emergency response measures aided in limiting the damaging effects of potential oil supply and price disruptions. The second characteristic has been *flexibility* in the formulation of energy policies. The IEA has endorsed a range of options as member countries have sought to develop sustainable energy policies and a continuation of this approach is likely to maximise the gains from future interactions with non-member countries.

In the process of developing energy strategies with its member countries, the IEA, as an institution, has developed a considerable body of knowledge and experience and it is likely to be in the dissemination of this expertise that the Agency will find its most appropriate role in dealing with non-member countries. Contacts of this nature have already been established with the Soviet Union and some of the economies of eastern Europe. In the case of the USSR, the Secretariat provided the energy input to the recently completed IMF/IBRD/OECD/EBRD study of the Soviet economy[1] in which recommendations were made for the restructuring of the energy sector. Future activities with the USSR will depend on the outcome of the current economic and political reform process and the speed with which energy sector reforms are undertaken. Given the potential importance of Soviet energy supplies to OECD economies, however, some priority will be attached to the maintenance of direct relations with this economy.

It is probably its relations with the eastern European economies, however, which will serve as a model for the IEA's future contacts with non-members. In response to the developments in the political situation in eastern Europe and their direct overtures for assistance, the IEA over the past year has developed a comprehensive programme for helping those countries to transform their energy sectors. The Agency has worked closely, in the G-24 context, with the EC Commission and the World Bank and its role as the principal western agency to evaluate and develop energy strategies in the region has been endorsed by these bodies. The in-depth country review process which has been developed in the member-country context will form the basis of initial contacts with non-members. It is expected that the IEA could also serve as a catalyst for the resolution of energy policy issues and problems through conferences, seminars and workshops.

Beyond the eastern European region, the IEA's role will probably be less formal and will depend upon the extent to which other non-

---

1. IMF, IBRD, OECD, EBRD (1991), *A Study of the Soviet Economy*. Paris.

members seek the Agency's assistance or advice. In terms of IEA priorities, the Asia-Pacific region will remain important because of its expected high rates of growth in energy demand, as will Latin America. Current efforts to increase direct contacts with these countries are to some extent constrained by the fact that most have not yet actively solicited IEA assistance. Recognising that mutual interests will be served by a sharing of knowledge and experience, the IEA will adopt a pro-active stance to the extent of participating in regional energy seminars and similar activities. In pursuing more direct links at an institutional level, however, a cautious and prudent approach will be required which recognises the economic and political circumstances of individual non-members.

The aftermath of the Gulf crisis has also provided opportunities for constructive discussions between the IEA and oil producing countries. These discussions recognise that expanded contacts between the principal oil consumers and producers are beneficial since they can promote increased market transparency, thereby enhancing its efficiency. Recognising that the market is the best allocator of resources, the proposed discussions will concentrate on technical areas including the promotion of energy data exchanges; industrial co-operation, dealing with issues such as investment requirements in the energy sector and foreign investment and fiscal regimes as well as the financing of energy projects; energy efficiency and environment; and market mechanisms.

The IEA will also continue a range of general activities which have implications for non-member countries. These include the updating and refining of its World Energy Balances database, which provides a comprehensive and consistent body of data for analysts and policy makers. It will continue its ongoing monitoring of energy and economic developments in non-member countries which provides an essential input to the energy demand forecasting process.

The Agency is also engaged in a number of actions to facilitate energy technology transfer and collaboration. Ongoing work in the area of energy research and development takes two basic forms. These are, first, the creation of greater awareness of the nature and scale of national research and development efforts and, second, the sponsoring of joint programmes and collaborative projects. Areas covered by the projects already concluded or underway include end-use efficiency, fossil fuels, renewables and nuclear fusion. Among its collaborative projects, the IEA has sponsored the establishment of two energy technology information systems, in recognition of the importance of disseminating the knowledge acquired from research and development. The Centre for the Analysis and Dissemination of Demonstrated Energy Technologies (CADDET) offers an international network to exchange information on demonstrations of energy-saving technologies for all consumers of energy. The Energy Technology Data Exchange (ETDE) is a central information-sharing system sponsored by a number of IEA countries. It maintains a database of over two million descriptions of energy research and development projects in the areas of fossil fuels, renewables, and energy storage and conversion.

The IEA is also playing a direct role in the international environmental arena and environmental issues have become an increasingly important component of its work programme. Most of this work is directed towards supporting member country participation in the negotiations towards a Global Climate Framework Convention and the United Nations conference on Environment and Development scheduled for 1992. Studies are being prepared, for example, on the potential for emissions reductions in the electricity sector as well as the potential of energy efficiency in general, and an analysis of the economic costs and social implications of the various policy options and instruments that are being considered to reduce global greenhouse gas emissions. The Agency is also developing an emission co-efficients database which could play a critical role in monitoring the outcome of any agreements reached in the international framework.

III. RECENT DEVELOPMENTS IN NON-OECD ENERGY SUPPLY AND PRODUCTION

PRIMARY ENERGY SUPPLY

Total Primary Energy

Developments in the level and structure of non-OECD energy demand and supply over the period 1973 to 1989 have been significant, although many variations have been evident between regions and countries. The non-OECD world as a whole consumed in 1989 almost twice as much energy as it did in 1973[1]. During the coming decade, non-OECD countries are expected to account for more than half of total world energy consumption, leaving the OECD region as the lesser player on the energy scene.

World total primary energy supply (TPES) — excluding traditional, non-commercial sources of energy such as fuelwood and biomass — increased by 44 per cent in the 16 year period between 1973 and 1989. In 1973, world energy demand was equal to 5416 Mtoe and grew at an average annual rate of 2.3 per cent to reach 7816 Mtoe in 1989.

Over the same period, TPES in the non-OECD countries almost doubled, growing at an average annual rate of 4.1 per cent, from 2005 Mtoe in 1973 to 3829 Mtoe in 1989. This was substantially higher than energy demand growth in the OECD region, which registered an average annual growth rate of 1.0 per cent. The rate of

---

1. The data on which this report is based are from IEA, *World Energy Statistics and Balances* for non-OECD countries and *Energy Balances of OECD Countries* for the OECD region, various years.

growth in non-OECD energy demand varied by region but was, in general, supported by high rates of population growth and increasing levels of industrialisation and private transportation. In addition, much less success was recorded in reducing energy intensities in the non-OECD world than in the OECD. As a result of these differential growth rates in energy demand, the non-OECD region increased its share of world total primary energy supply. From 37 per cent in 1973, its share had risen to 49 per cent in 1989.

Figure 1: **SHARES OF WORLD TPES**

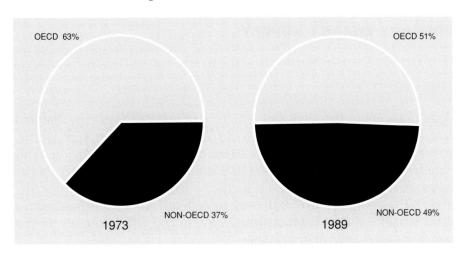

Source: IEA Secretariat

Among the non-OECD regions, the USSR and China are the largest absolute consumers of energy. When eastern Europe is included, their combined share of non-OECD TPES was 65 per cent in 1989. In the USSR, total energy demand grew at 3.1 per cent per year between 1973 and 1989. As a result of deteriorating economic performance, however, growth was considerably lower in the 1980s than in the 1970s. In eastern Europe, TPES growth averaged 2.0 per cent per year, lower than in any other non-OECD region. In China, energy demand increased at a rate of 5.8 per cent per year, reflecting high rates of growth in both population and industrial output. While total energy demand growth in these combined regions was the lowest in the non-OECD area, it is worth noting that it was about three times higher than that of the OECD over the same period. All other regions increased their shares of non-OECD TPES to varying degrees. In 1989, 13 per cent of the total was accounted for by the Asia-Pacific region, 10 per cent by Latin America, and 6 per cent by each of the Middle East and Africa. The highest TPES growth rates among this group of regions were experienced by the Middle East and the Asia-Pacific region, followed by Africa and Latin America.

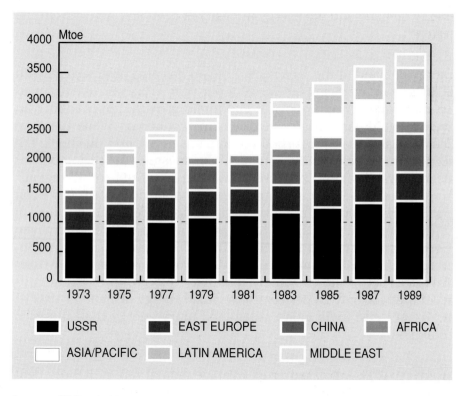

Figure 2: **NON-OECD TPES BY REGION**

Source: IEA Secretariat

Energy demand growth in the Middle East has been especially strong, averaging 8.6 per cent per year over the review period. In both the 1970s and the 1980s growth was the highest of any non-OECD region. Energy demand was supported, in part, by large increases in population — a number of countries in the region recorded population growth rates in excess of 5 per cent per year over the period. A further important factor was the large-scale development of an energy-intensive, heavy and chemical industrial sector in the Middle East. A number of the Gulf states have used their enormous oil revenues to pursue policies aimed at diversifying their industrial structures away from dependence on crude oil exports. As a result, a range of industries, including refining, petrochemicals and aluminium smelting, has emerged. This experience of industrialisation has been very different from the labour-intensive, export-oriented policies pursued in many of the developing Asian economies and has had implications for total energy demand and the energy intensity of economic activities in the region.

Energy demand in the Asia-Pacific region has also been strong, with TPES growth averaging 6.2 per cent per year. This was underpinned by particularly high levels of economic growth, approaching an average of 10 per cent per year in some countries. While the Asia-Pacific region easily absorbed the impacts of the first oil shock, the second crisis did slow the economic dynamism of the region but for a considerably shorter period than was experienced in other regions. Among the Asia-Pacific countries, energy demand growth has been particularly rapid in the newly industrialising economies (NIEs) of South Korea, Taiwan, Hong Kong and Singapore. TPES increased 3.4 times in this group of countries between 1973 and 1989, compared with 2.6 times for the Asia-Pacific region as a whole. Nevertheless, while extremely dynamic, this group of economies accounted for only 28 per cent of regional energy demand in 1989 — at 34 per cent, India alone consumed more energy than the NIEs combined.

In Africa, TPES increased at an average annual rate of 5.8 per cent over the review period, but a notable feature of the region is that only a small group of countries account for the major proportion of regional energy demand. This has become more pronounced over time. Among about 50 countries in Africa, Algeria, Egypt, Libya, Nigeria and South Africa combined accounted for 72 per cent of regional TPES in 1973 and 82 per cent in 1989. While most of the sub-Saharan countries (including Nigeria) account for about 60 per cent of the regional population, their energy demand amounted to only 26 per cent of the total in 1973 and 20 per cent in 1989. Energy demand in the sub-Sahara has remained low because of its generally low levels of economic development, reflected particularly in its industrialisation and urbanisation rates and its levels of per capita income, as well as by the continuing substantial reliance on traditional fuels in many countries.

TPES in Latin America grew at an average annual rate of 3.8 per cent over the period under review, slower than all other non-OECD regions except the USSR and eastern Europe. Growth was much lower during the 1980s than the 1970s, reflecting the general economic problems of many countries in the region.

The share of each fuel in the non-OECD's TPES has also changed over time, with increased use of natural gas and decreased dependence on coal and oil. In 1973, coal and oil together accounted for 83 per cent of total primary energy supply, but by 1989 their combined share had fallen to 72 per cent. Most of the capacity was taken by gas, which increased its share from 15 per cent to 23 per cent over the same period. At the same time as these developments were

Figure 3: **NON-OECD TPES BY FUEL**

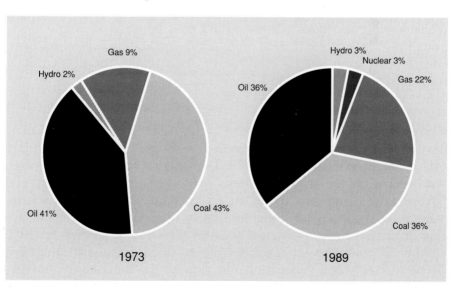

Source: IEA Secretariat

GLOBAL ENERGY

occurring, the OECD region's dependence on oil was also declining but the shortfall was made up by nuclear power and, to a lesser extent, coal.

In the USSR and eastern Europe, the share of coal and, to a lesser degree, that of oil, in total TPES has declined over time. The share of gas rose sharply, especially in the USSR where it accounted for over 40 per cent of TPES in 1989, compared with 32 per cent for oil and 22 per cent for coal. Coal has remained the principal fuel in the eastern European demand structure and, in both regions, nuclear's role was still small in relative terms in 1989. In the Asia-Pacific region, there was an increased dependence on gas and nuclear power, and renewed emphasis on coal in the 1980s, while dependence on oil declined. In Africa, fuel demand continued to be dominated by coal and oil, although gas increased its contribution to TPES over time. Hydropower increased its share of TPES in the Latin American region, as did gas, with both fuels replacing the declining share of oil. In the Middle East, oil and gas combined accounted for the larger share of TPES over the study period, although there was an increasing emphasis on gas.

Coal

While world coal demand grew 50 per cent between 1973 and 1989, from 1567 Mtoe to 2357 Mtoe, non-OECD demand increased by over 60 per cent, from 852 to 1385 Mtoe. As a result, the non-OECD's share of world coal demand increased from 54 per cent to 59 per cent.

Among the non-OECD regions, the USSR, eastern Europe and China have been by far the largest consumers of coal, representing 78 per cent of the non-OECD total in 1989. Demand for coal in these regions has been based on large indigenous reserves and, at least in eastern Europe and China, has been particularly important in the power generation sector. Coal consumption in the Asia-Pacific region has also grown strongly, particularly during the 1980s, as policies in a number of countries have been directed towards substituting coal for oil in electricity generation. In view of the increasing concern given to the environmental consequences of coal use, it is uncertain for how long this trend will be sustained.

Figure 4: **SHARES OF WORLD COAL DEMAND**

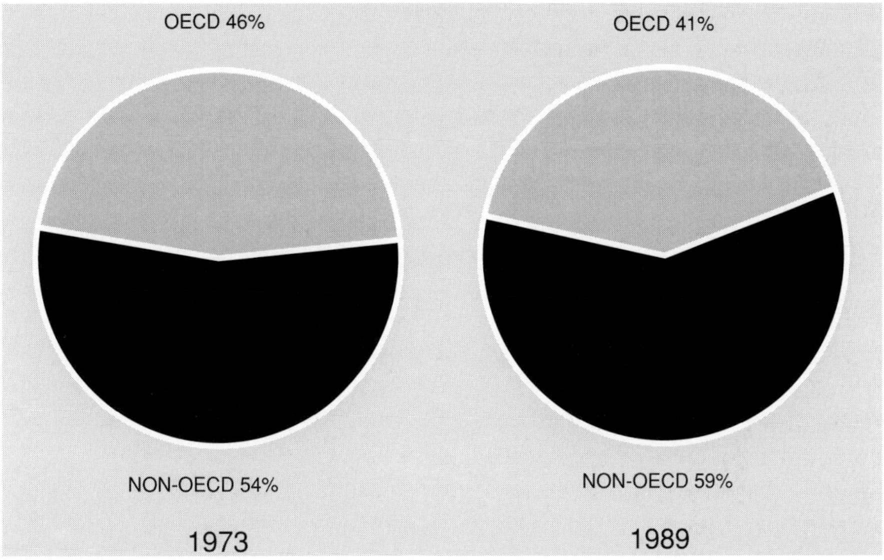

1973	1989
OECD 46%	OECD 41%
NON-OECD 54%	NON-OECD 59%

Source: IEA Secretariat

Oil

Total world oil demand increased 15 per cent between 1973 and 1989, from 2700 Mtoe to 3093 Mtoe, all of which was accounted for by increases in the non-OECD world. OECD oil demand was lower in 1989 than in 1973, despite increases in some intervening years. In contrast, oil demand in the non-OECD continued to increase between 1973 and 1989, from 815 Mtoe to 1367 Mtoe. Growth continued even during the period following the second oil shock, although the growth rates experienced in the 1980s were substantially below those of the 1970s. Accordingly, the non-OECD's share of world oil demand rose to 44 per cent in 1989, from 30 per cent in 1973.

Among the non-OECD regions, oil demand growth remained strongest in the Middle East, followed by Africa, China and the Asia-Pacific. Demand was weakest in the USSR and eastern Europe, but in absolute terms the USSR continued to dominate non-OECD oil demand, accounting for almost one-third of the total in 1989.

Figure 5: **SHARES OF WORLD OIL DEMAND**

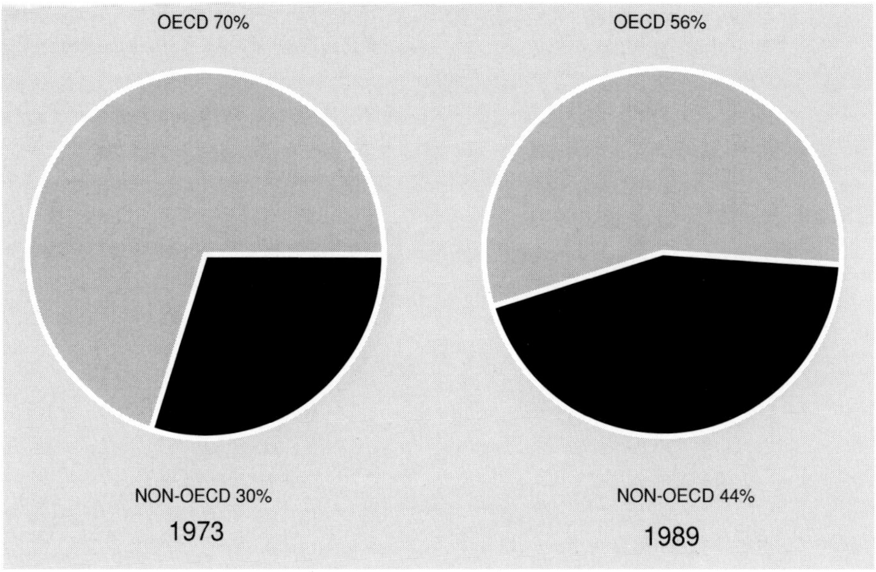

OECD 70%	OECD 56%
NON-OECD 30%	NON-OECD 44%
1973	**1989**

Source: IEA Secretariat

Natural Gas

World natural gas demand increased 69 per cent over the study period, from 976 Mtoe in 1973 to 1652 Mtoe in 1989. The growth in non-OECD natural gas demand was particularly strong. Between 1973 and 1989, non-OECD demand almost tripled, from 297 Mtoe to 878 Mtoe, while demand in the OECD, already a relatively mature gas market, increased by only 14 per cent. OECD demand actually declined in volume terms during the early 1980s. Accordingly, the non-OECD now consumes more than half the world's natural gas supply, compared with 30 per cent in 1973.

Within the non-OECD, the USSR accounts for about two-thirds of total natural gas demand. Growth rates in the Asia-Pacific and Middle East regions, however, were especially rapid during the 1980s, and their share of total non-OECD demand rose. These developments largely reflect the introduction of natural gas for electricity generation and an increased range of industrial applications, releasing oil for export in the producer regions.

Figure 6: **SHARES OF WORLD GAS DEMAND**

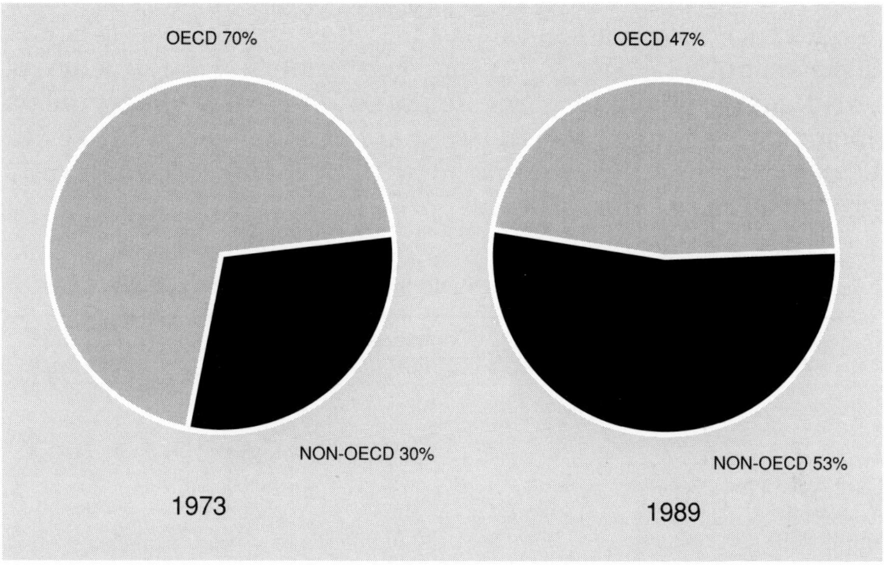

1973	1989
OECD 70%	OECD 47%
NON-OECD 30%	NON-OECD 53%

Source: IEA Secretariat

Traditional Fuels

While the above data refer only to commercial sources of energy, in some regions of the world traditional fuels make a significant contribution to total energy production and consumption. Traditional fuels are defined to include fuelwood, bagasse (the cellulosic residue left after sugar is extracted from sugar cane), charcoal, animal and vegetal wastes and peat. To a large extent these are non-commercial, i.e. non-traded, fuel sources but in some cases may act as substitutes for commercial fuels, for example, where bagasse is used as a fuel in the sugar milling industry. The non-traded nature of many traditional fuels makes the measurement of their production and consumption more difficult than that of commercial fuels. Hence, the coverage of traditional fuel statistics is likely to be less complete and to underestimate the true extent of their use.

Table 1 outlines the dimensions of traditional fuel demand in non-OECD regions and compares this with commercial energy demand. As a share of total energy demand, traditional fuels have been most important in Africa, where they accounted for 33 per cent in 1989, Asia-Pacific (24 per cent), Latin America (19 per cent) and China (6 per cent). For the non-OECD as a whole, traditional fuels met 10 per cent of total energy demand. Furthermore, if the structure of world energy demand is recalculated on a total energy basis (including traditional fuels), the share of the non-OECD rises to 52 per cent in 1989, compared with 49 per cent when only commercial fuels are considered.

Table 1: **Traditional Fuel Demand in Non-OECD Regions, 1989**

	Traditional Fuel	Commercial Fuel	Total Fuel	Traditional/ Total Fuel
	Mtoe	Mtoe	Mtoe	%
Africa	107	220	327	33
Latin America	88	376	464	19
Asia-Pacific	159	509	668	24
Middle East	1	230	231	—
Eastern Europe	6	480	486	1
USSR	25	1362	1387	2
China	43	650	693	6
Total non-OECD	**429**	**3827**	**4256**	**10**

Source: IEA Secretariat

In all the non-OECD regions where traditional fuels are important, their share of total energy demand has fallen steadily since 1973, reflecting the shift into commercial fuels that generally accompanies rising levels of economic development and per capita incomes. This shift is expected to continue and will have implications for the future level and structure of energy consumption.

PRIMARY ENERGY PRODUCTION

Total Primary Energy Production

World total primary energy production increased 41 per cent in the 16 years between 1973 and 1989, from 5616 Mtoe to 7900 Mtoe. This represented an average annual growth rate of 2.2 per cent and built on already substantial production gains achieved in the 1960s. Over the same period, non-OECD energy production increased 45 per cent, from 3462 Mtoe to 5005 Mtoe. It grew at an annual rate of 2.3 per cent, compared with 1.9 per cent in OECD countries. The non-OECD share of world primary energy production was 63 per cent in 1989, little changed from its position in 1973.

Figure 7: **SHARES OF WORLD PRIMARY ENERGY PRODUCTION**

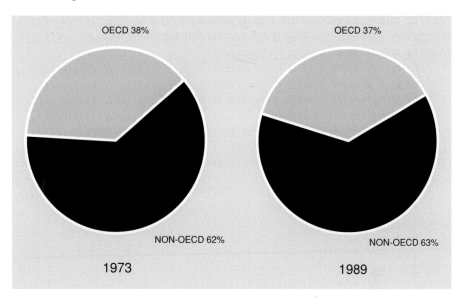

Source: IEA Secretariat

The USSR emerged during the 1980s as the world's largest primary energy producer, accounting for about one-fifth of the world total and one-third of the non-OECD total in 1989. This position, however, has not been achieved without cost. The USSR's output of natural gas has increased continuously, but production in the coal sector stagnated in the 1980s and oil production has declined in the latter part of the decade. Although nuclear power production has risen sharply, it has failed to achieve targets and the Chernobyl incident has seriously affected the future of the nuclear programme. Continued investment, including the introduction of new technology and equipment, will be necessary if energy production levels are to be sustained.

Although its total output declined over the review period, the Middle East remained the second largest primary energy producer in the non-OECD region, followed by China, Latin America and Africa. In Africa, primary energy production increased 35 per cent between 1973 and 1989 — lower than the non-OECD average. Four countries in the region, Algeria, Egypt, Nigeria and South Africa, increased their combined share of the region's output from 60 to 73 per cent over the period, leading to the conclusion that in most other African countries energy production probably deteriorated. The Asia-Pacific region recorded rapid growth in energy production, principally natural gas, but it remained the smallest overall producer, with the exception of eastern Europe.

Figure 8: **NON-OECD ENERGY PRODUCTION BY REGION**

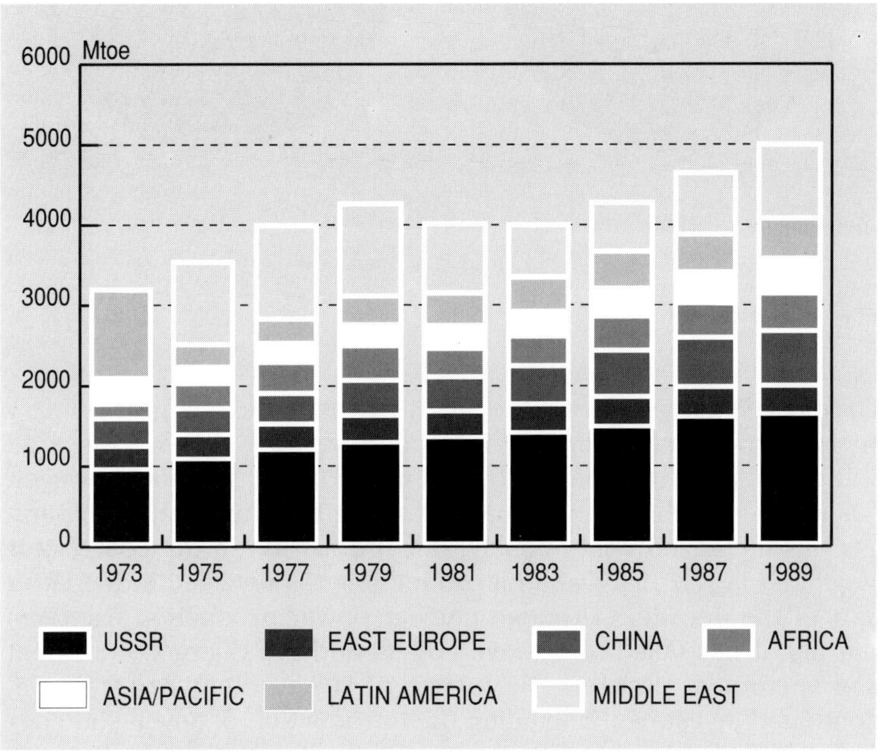

Source: IEA Secretariat

The overall structure of energy production in the non-OECD has become increasingly weighted towards natural gas, with the relative importance of oil declining over time. In 1973, oil accounted for two-thirds of total non-OECD primary energy output, while gas accounted for only 9 per cent. By 1989, however, oil's share had fallen to 49 per cent and that of gas had increased to 20 per cent. Coal's share remained fairly constant over the period at close to one-quarter.

Figure 9: **NON-OECD ENERGY PRODUCTION BY FUEL**

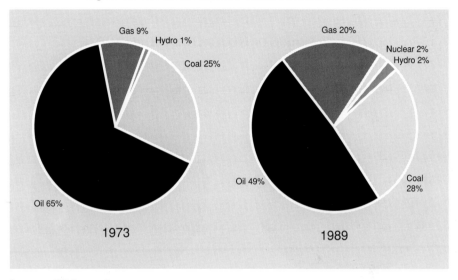

Source: IEA Secretariat

These changes in production structure varied considerably across regions. In the USSR, the shares of coal and oil in total primary energy production declined, with sharp increases in natural gas output. The changes in eastern Europe were less significant, with coal maintaining over three-quarters of total production. China's production structure also remained stable, with about 75 per cent coal and 20 per cent oil. In Asia-Pacific, the shares of gas and nuclear power rose at the expense of coal and oil, while in Africa, increased weight was placed on coal and natural gas and oil's share of the total fell. In Latin America, the share of hydropower increased as well as that of gas and the contribution of coal and nuclear electricity remained small. In the Middle East, virtually all primary energy production was of oil and natural gas, with gas increasing its share over time.

Coal

World production of coal increased about 50 per cent over the review period, from 1545 Mtoe in 1973 to 2339 Mtoe in 1989. At the same time, non-OECD coal production increased 60 per cent, from 867 Mtoe to 1385 Mtoe, while the OECD region exhibited a 41 per cent increase. The non-OECD's share of world coal production rose, therefore, from 56 per cent in 1973 to 59 per cent in 1989.

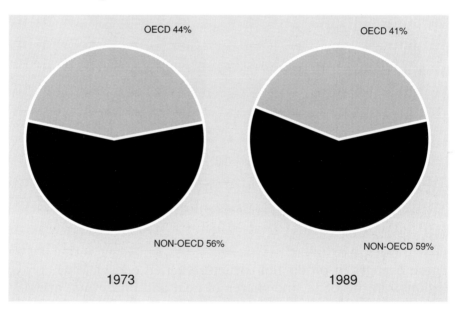

Figure 10: **SHARES OF WORLD COAL PRODUCTION**

Source: IEA Secretariat

Together, the USSR, eastern Europe and China have been the largest producers of coal, accounting for 80 per cent to 90 per cent of the non-OECD total over the review period. China has emerged as the second largest producer, as well as consumer, of coal in the world after the United States. The highest growth in coal production, however, was recorded by Latin America, although the region accounted for less than 2 per cent of the non-OECD total in 1989.

Oil

World oil production generally fluctuated over the period studied, rising until the impacts of the second oil shock were felt on demand. The recovery from that period commenced in 1984 and by 1989 production was equal to 3189 Mtoe, 10 per cent higher than in 1973. Non-OECD production followed a similar path, peaking in 1979, followed by a decline through 1983. Production in 1989 was 2429 Mtoe, 8 per cent higher than the 1973 level of 2247 Mtoe. The non-OECD's share of world oil production changed accordingly over the period; from 77 per cent in 1973, it reached its peak of 79 per cent in the late 1970s and was 76 per cent in 1989.

Figure 11: **SHARES OF WORLD OIL PRODUCTION**

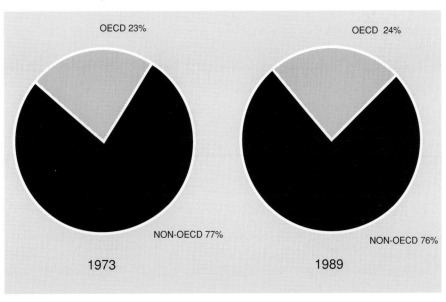

Source: IEA Secretariat

Among the non-OECD regions, Middle East production levels varied considerably over the period and in 1989 were nearly 22 per cent lower than in 1973. In the USSR, production increased steadily until the mid-1980s. Despite recent declines, the USSR has remained the world's largest single producer of oil, accounting for 19 per cent of the total in 1989. This compares with 26 per cent for the Middle East in the same year.

Natural Gas

World production of natural gas rose 67 per cent over the review period, from 990 Mtoe in 1973 to 1657 Mtoe in 1989. Growth in non-OECD production was rapid, with total output more than tripling, from 307 Mtoe in 1973 to 992 Mtoe in 1989. At the same time, OECD production of natural gas declined for a period after 1979 and has recovered only slowly from its lowest level in 1983. As a result, the non-OECD's share of the world total almost doubled, from 31 per cent in 1973 to 60 per cent in 1989. The USSR has been by far the world's largest producer of natural gas, accounting for about 65 per cent of the non-OECD total and 39 per cent of the world total in 1989.

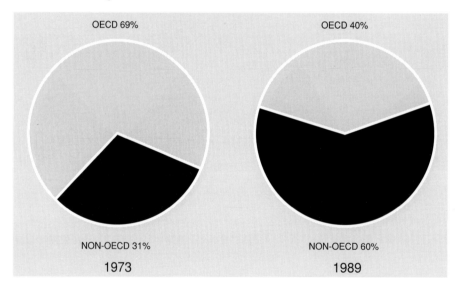

Figure 12: **SHARES OF WORLD GAS PRODUCTION**

Source: IEA Secretariat

Nuclear Power

World output of nuclear electricity increased almost 10 times between 1973 and 1989, from 53 Mtoe to 506 Mtoe, with most of the growth in absolute terms occurring in the OECD region. Indeed, in global terms nuclear power capacity remains heavily concentrated in OECD countries. Non-OECD nuclear production increased from

4 Mtoe to 100 Mtoe, while OECD output rose from 49 Mtoe to 406 Mtoe. As a consequence, the non-OECD share of world nuclear output rose from 8 per cent in 1973 to 20 per cent in 1989.

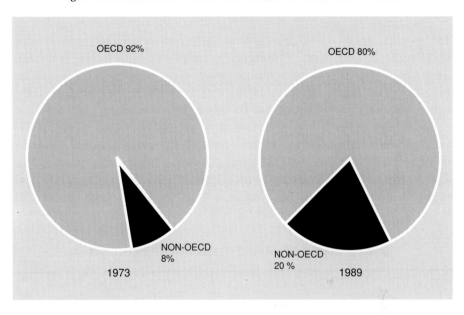

Figure 13: **SHARES OF WORLD NUCLEAR POWER PRODUCTION**

Source: IEA Secretariat

Within the non-OECD region, the USSR, eastern Europe and Asia-Pacific were the only producers of nuclear power until the mid-1970s, with the USSR alone accounting for three-quarters of the total in 1973. Latin America's first nuclear plant came on stream in the mid-1970s, followed by Africa in the mid-1980s. No nuclear capacity had been installed in the Middle East or in China by 1989. The strong penetration of nuclear power in South Korea and Taiwan produced growth rates in the Asia-Pacific region which were higher than elsewhere in the non-OECD, thereby increasing Asia-Pacific's share of total non-OECD output to 22 per cent by 1989. In the same year, the USSR accounted for 55 per cent.

Hydropower and Other Renewables

World output of electricity from hydro and other sources (geothermal, solar, wind and other renewable fuels) increased 74 per cent in the 1973-1989 period, from 119 Mtoe to 207 Mtoe. The overwhelming majority of this output was attributable to hydropower sources. Only in Latin America, Africa and the Asia-Pacific region did other renewable forms of energy make any measurable contribution to electricity output. In Africa, the share of other renewables in "hydro and other" was 6 per cent, in Latin America 13 per cent and in Asia-Pacific 25 per cent. The figure is relatively high in the last case because of the overall insignificance of hydropower in total energy output. (Given the dominance of hydro in these regional figures, future reference will be limited to this form of energy, although the data in fact refer to hydro and other renewables). Non-OECD hydropower output almost tripled over the period, from 37 Mtoe to 98 Mtoe, compared with 33 per cent growth in OECD countries where a large proportion of hydro potential had already been exploited. The non-OECD share of world hydroelectricity production rose from 31 per cent in 1973 to 47 per cent in 1989.

Figure 14: **SHARES OF WORLD HYDROPOWER PRODUCTION**

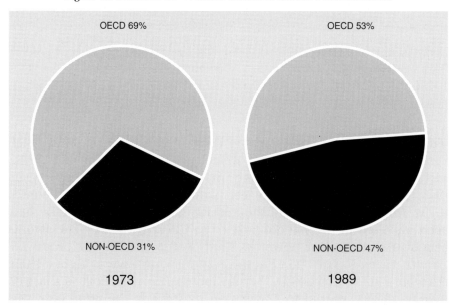

Source: IEA Secretariat

Among the non-OECD regions, Latin America has been the largest producer of hydropower and its share of the total rose from 27 per cent in 1973 to 38 per cent in 1989. The USSR has been the second largest producer over the period, although its share fell from 32 per cent to 21 per cent. Other areas to gain in share were the Asia-Pacific region and China. Production of hydropower in the African region fell during the early 1980s as a result of widespread and sustained drought conditions, bringing the region's share of the non-OECD total to 5 per cent in 1989 compared with 8 per cent in 1973.

ELECTRICITY GENERATION

Electricity generation more than doubled in the non-OECD region, from 1959 TWh in 1973 to 4767 TWh in 1989. By comparison, OECD electricity generation increased 59 per cent over the same period and hence the non-OECD's share of world electricity output rose to 42 per cent in 1989 from 32 per cent in 1973. The rapid growth in non-OECD electricity demand has been driven by a growing need for energy services most easily provided by electricity such as lighting, heating and, to a lesser extent, cooking. The opportunities for increased electricity use have expanded as economic and income growth have continued, bringing with them increased demand for quality of life improvements. This is reflected in the fact that electricity demand growth in the non-OECD has been greater in the household and commercial sectors than in the industry sector.

Overall, coal has remained the dominant fuel for electricity generation purposes in the non-OECD region, but its share of total inputs declined from 48 per cent to 41 per cent over the study period. On a regional basis, coal has remained dominant in eastern Europe and Africa, although its share of total inputs declined in both regions. In China, coal's share of inputs increased to 82 per cent. In the Asia-Pacific region, coal inputs almost tripled during the 1980s, with coal replacing oil as the principal fuel for electricity generation.

The major gains in non-OECD fuel share were made by natural gas and nuclear power, whose shares rose from 14 to 23 per cent and from virtually zero to 10 per cent respectively. While the gains in natural gas were made across all regions except Latin America, the increase in nuclear's share was confined largely to the USSR, eastern Europe and the Asia-Pacific region.

Figure 15: **INPUTS TO ELECTRICITY GENERATION**

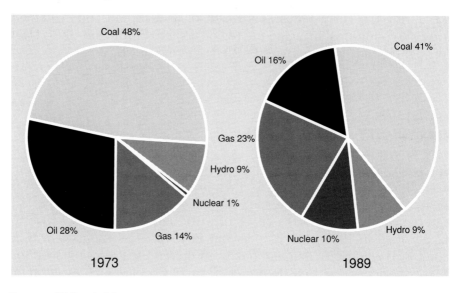

Source: IEA Secretariat

Oil-fired electricity generation increased across the non-OECD throughout the 1970s, to a peak of 25 per cent in 1980, but declined thereafter to 16 per cent in 1989. Reflecting natural endowments, the share of oil in the Middle East's electricity generation has never fallen below about 50 per cent, but was about 13 per cent in 1989 in the USSR, 19 per cent in the Asia-Pacific region and 22 per cent in Latin America. This compares with a figure of 10 per cent for the OECD.

ENERGY INTENSITY

A number of measures aimed at energy conservation and improvements in the efficiency of energy use were taken in some non-OECD countries over the period under review. But in terms of energy intensity — measured as the ratio of TPES (in toe) to GDP (in $US1000 at 1985 prices and exchange rates) — the situation in general did not improve; i.e. in many countries, more energy was required to produce a given unit of GDP in 1989 than in 1973. This contrasts with the experience in OECD countries over the same period where energy intensities fell by an average of 1.7 per cent per year. The trends in the non-OECD region reflect a variety of factors associated with rising income levels, including the processes of industrialisation and urbanisation and increased demand for transport fuels. They often reflect, in addition, changes in the structure of an economy's output towards more energy-intensive activities. In a range of countries, the trend has been compounded

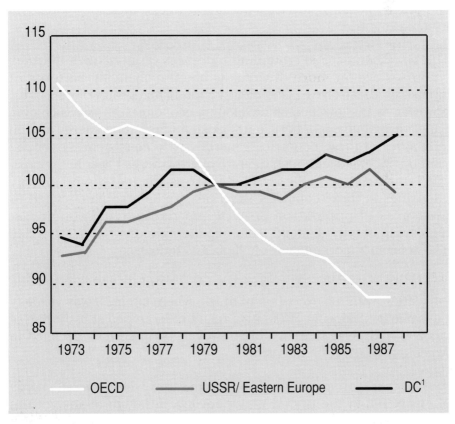

Figure 16: **ENERGY INTENSITY (TPES/GDP)**
1980 = 100

1. Developing countries
Source: IEA Secretariat

by inefficient energy pricing policies which have discouraged energy conservation measures.

Table 2 shows that the countries with the highest rates of growth in energy intensity have been the energy resource-rich economies of the Middle East, as well as Algeria and Nigeria. In Saudi Arabia, the UAE and Kuwait, for example, total energy intensity increased by more than 7 per cent per year, and in the African economies the rate was in excess of 6 per cent. In the Middle East economies, the rate of increase in oil intensity was generally higher than that for total energy, reaching 13 per cent per year in Saudi Arabia and almost 15 per cent per year in the UAE. The pressures for increased intensities in the Middle East and Algeria were clearly related to structural transformations favouring the emergence of energy-intensive, oil- and gas-based industrial sectors. Government industry and pricing policies played an important role in these developments in some countries.

The circumstances surrounding increases in intensities in other countries are far more diverse, as are the changing patterns of intensities themselves. One clearly identifiable group of countries, however, is the low-income developing economies of Asia, including Bangladesh, India, Indonesia and Pakistan. Both total energy and oil intensities in these countries generally rose consistently over the period concerned, although the rate of growth was lower in the post-1985 period than earlier. The reasons for increasing energy intensities in these countries vary, but are in large part related to pressures of population growth and rising living standards. Increasing energy demand from expanding industrial sectors has also been an important factor in India and Indonesia.

The economies of Latin America also exhibited rising energy intensities, with the exception of Brazil, where intensity was virtually the same in 1989 as in 1973. As a result of increased substitution of gas and hydroelectricity for oil, however, oil intensities declined in Argentina, Brazil and Colombia.

At the other extreme is the group of countries where overall energy intensity fell in the period under review, although in several the decline was limited to the 1980s. This group includes most of the dynamic Asian economies (Hong Kong, Singapore, South Korea,

Taiwan and Thailand). Oil intensities also generally fell over the same period in these economies. In these cases, efficiency improvements can generally be associated with structural transformations. In Hong Kong, for example, the share of services in total GDP increased at the expense of industry between 1973 and 1989. In South Korea, increases in energy intensity during the 1970s were associated with a period of rapid, heavy industrial development which was followed in the 1980s by diversification into lighter industry and the services sector.

Because of the potential impact of non-commercial fuel consumption on measures of energy intensity, the above relationships have been recalculated using the ratio of all energy consumption, i.e. commercial and non-commercial, to GDP. Two major points emerge. The first and most expected is that the absolute level of energy intensity is, in certain countries, considerably higher than when commercial energy only is considered. More interesting, however, is that the rate of growth of energy intensity is generally lower when non-commercial energy consumption is included in the analysis. The reason for the difference is that with increasing levels of development and income there is an observable shift in energy consumption away from non-commercial to commercial fuels. To this extent, a measure which includes only commercial fuels as its yardstick of energy consumption is measuring a broader base — or a higher proportion of total consumption — in each consecutive year. Hence, an analysis based only on commercial fuels will tend to overstate the deterioration in developing country energy intensities between 1973 and 1989 and to widen the apparent gap between OECD and non-OECD experience. In reality, the historical path of non-OECD energy intensities has not been as disappointing as commercial fuel analysis alone would indicate. The shift to higher proportions of commercial fuels in the total energy balance will, in fact, act in the longer term to lower energy intensities as commercial fuels are generally more efficient than non-commercial. The average efficiency of a fuel-wood stove, for example, is about 12 to 18 per cent, while the efficiency of a simple LPG-fired stove can be as high as 50 per cent.

Many other factors will compete, however, with this tendency towards lower intensities, chief among them being the inevitable pressure of population growth. As Table 2 indicates, per capita energy intensities are still remarkably low in many non-OECD countries, particularly those of Africa and developing Asia. In 1989, the average per capita energy intensity of OECD countries was 4.8 toe per person and, since 1973, had grown at an average annual rate of only 0.2 per cent. In the lowest-income developing countries, per capita intensities ranged from 0.06 toe per person in Bangladesh

Table 2: **Energy Intensity — Selected Countries, 1973 and 1989**

	1973		1989		Average Annual Growth %	
	TPES/GDP[1]	TPES/POP[2]	TPES/GDP[1]	TPES/POP[2]	TPES/GDP[1]	TPES/POP[2]
Africa						
Algeria	0.16	0.30	0.45	1.08	6.7	8.3
Egypt	0.44	0.21	0.58	0.55	1.7	6.2
Nigeria	0.04	0.05	0.16	0.14	9.1	6.6
South Africa	1.20	1.94	1.75	2.81	2.4	2.3
Asia-Pacific						
Bangladesh	0.18	0.02	0.36	0.06	4.4	7.1
Hong Kong	0.28	0.88	0.25	1.90	-0.7	4.9
India	0.58	0.12	0.71	0.21	1.3	3.6
Indonesia	0.26	0.09	0.38	0.23	2.4	6.0
Malaysia	0.36	0.43	0.50	1.01	2.1	5.5
Pakistan	0.53	0.11	0.64	0.21	1.2	4.1
Philippines	0.48	0.25	0.48	0.29	0.0	0.9
Singapore	0.56	1.86	0.43	3.50	-1.6	4.0
South Korea	0.65	0.64	0.65	1.89	0.0	7.0
Taiwan	0.51	0.84	0.50[3]	2.19	-0.1	6.2
Thailand	0.54	0.22	0.47	0.44	-0.9	4.4
Latin America						
Argentina	0.60	1.34	0.77	1.43	1.6	0.4
Brazil	0.41	0.47	0.40	0.66	-0.2	2.1
Colombia	0.52	0.49	0.52	0.61	0.0	1.4
Mexico	0.43	0.79	0.67	1.36	2.8	3.5
Venezuela	0.44	2.10	0.62	2.07	2.2	-0.1
Middle East						
Iran	0.17	0.78	0.40[3]	1.15	5.5	2.5
Kuwait	0.18	6.16	0.55	6.10	7.2	-0.1
Saudi Arabia	0.10	1.01	0.67	4.69	12.6	10.1
UAE	0.13	3.53	0.70	14.40	11.1	9.2
OECD	**0.52**	**4.62**	**0.40**	**4.79**	**-1.7**	**0.2**

1 Metric tons of oil equivalent per $1000 of GDP at 1985 prices and exchange rates
2 Metric tons of oil equivalent per inhabitant
3 Latest available data are 1988

Source: IEA Secretariat

to 0.1 in Nigeria and 0.2 in India, Indonesia and Pakistan. In general, and given a range of other competing variables, per capita intensities tend to increase as the level of income and development increases. Hence, per capita intensities in the newly industrialising economies of Asia are considerably higher than those in their lower-income neighbours and include 1.89 in South Korea, 1.90 in Hong Kong, 2.19 in Taiwan and 3.50 in Singapore. Per capita intensities approaching or exceeding those of OECD countries were experienced in the developed Middle East economies of Saudi Arabia (4.69), Kuwait (6.10) and the UAE (14.40).

While absolute levels remain low, the rate of growth in per capita intensities in non-OECD countries has significantly outpaced that in the OECD and it is unlikely that this trend will change. Enormous potential, especially in the developing world but also in the Soviet Union and eastern Europe, exists for increased energy consumption at the household level. Increased rural electrification and the accessibility to energy-consuming appliances associated with urbanisation and rising income levels will act to maintain or increase per capita intensities. This effect will be intensified by rising levels of personal transportation, accelerating industrialisation and high rates of population growth.

THE NON-OECD'S EXPORTABLE ENERGY SURPLUS — THE TRADE RELATIONSHIP WITH OECD COUNTRIES

As a result of the differential growth rates between non-OECD energy production and demand, the region's exportable energy surplus has shrunk considerably over time. An exportable energy surplus is defined as a region's excess of exports over imports and represents its net additions to the world market. When the non-OECD is considered as one region, its net exports are those available to the OECD. In 1973, 1378 Mtoe, equivalent to 69 per cent of the non-OECD's TPES, was available for export to OECD countries, whereas in 1989 the figure had shrunk to 1116 Mtoe or 29 per cent of TPES. This decline in the non-OECD's export performance was most pronounced between 1979 and 1983, reflecting the price-induced fall in global oil consumption over that period. In terms of the OECD's energy demand, the non-OECD's exportable energy surplus was equal to 40 per cent of TPES in 1973 and 28 per cent in 1989. The decline in the dependence of the OECD on the non-OECD implied by these figures reflects a combination of influences, including the much higher growth rates of energy consumption experienced by the non-OECD than the OECD, the greater success of the OECD in reducing energy intensities through structural change and improvements in energy efficiency, and the OECD's increases in indigenous energy production.

Figure 17: **NET ENERGY TRADE**

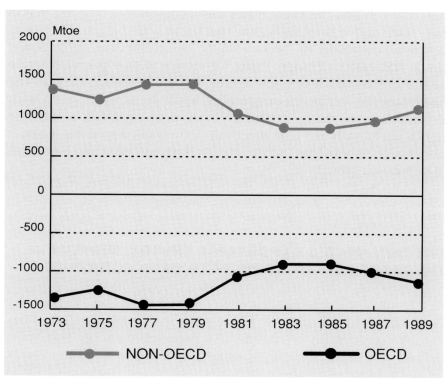

Source: IEA Secretariat

Of the non-OECD's total exportable surplus, the major share has been accounted for by oil — 98 per cent of the total in 1973 and 90 per cent in 1989. This surplus was equivalent to 72 per cent of OECD oil demand in 1973 and 58 per cent in 1989. The Middle East has been the major oil exporter, accounting for almost three-quarters of non-OECD net exports in 1973 and two-thirds in 1989. In that year the Middle East alone accounted for 44 per cent of OECD oil imports from the non-OECD, representing 28 per cent of OECD oil consumption. Africa and Latin America were the next most important suppliers of oil to the OECD, accounting for 13 per cent and 10 per cent, respectively, of total OECD consumption.

The availability of non-OECD natural gas for export has grown rapidly over time, increasing 13-fold between 1973 and 1989, although from a low base. While its contribution to the total non-OECD surplus was negligible in 1973, it had risen to 9 per cent by 1989 and was equivalent to 13 per cent of OECD gas consumption in that year. Most of the region's net exports of gas in 1989 were sourced from the USSR, Asia-Pacific and Africa, which were also the most important sources of supply to OECD countries. Imports from the USSR were equal to 5 per cent of OECD gas consumption in 1989, and those from the Asia-Pacific region to 4 per cent.

With abundant and geographically dispersed coal reserves, the OECD as a region has been relatively self-sufficient in coal. As a result of the wide global differences in coal qualities and applications, however, there has been a substantial amount of coal trade between OECD and non-OECD countries. The non-OECD coal surplus was 8 Mtoe in 1989, the majority of which was sourced from Africa — principally South Africa — and the USSR.

IV. THE ENERGY OUTLOOK IN NON-OECD COUNTRIES

INTRODUCTION

The projections of energy demand and supply contained in this section are based on the IEA's "World Energy Outlook". The outlook presented here is intended to provide only a general indication of the direction and possible magnitude of changes in worldwide energy demand and production. There are two areas, in particular, which are likely to increase the uncertainty of the forecasts. The first of these concerns the rapid political transformation of eastern and central Europe and the former Soviet Union and the possibilities this provides for economic changes and, consequently, changes in the energy economy of these regions. The second area concerns the relationship between energy consumption and environmental protection. The projections assume that current government energy policies and current trends in environmental protection continue unchanged. However, if environmental concerns were to lead to major technological breakthroughs or if environmental policies resulted in aggressive promotion of energy efficiency and fuel substitution, the implications for global energy demand and supply could be very significant. Many OECD governments, for example, are currently contemplating far-reaching changes to stem the growth of energy-related greenhouse gas emissions. Some of these changes, if adopted, could have a significant impact on how energy is produced and consumed and, hence, on the energy markets facing non-member countries as well.

As with any energy outlook, the reference case from which the projections in this report are taken is based on a number of different assumptions, principally regarding crude oil prices and economic growth. Specifically, the outlook assumes that crude oil prices (in constant 1990 United States dollars) settle at around $21 per barrel in 1992 and then rise gradually to about $35 by early in the next decade and remain at that level thereafter. The oil price underlying this scenario is clearly only one of a number of possible price paths. It assumes that crude oil prices, in real terms, begin to rise as excess production capacity outside the Gulf-producing region is reduced. As this occurs, supply and price will depend upon the extent to which Middle East production and, ultimately, capacity are increased. This particular scenario assumes that Middle East oil production is expanded in the 1990s at a rate which also permits some real price increases to take effect. A price of $35 per barrel is roughly equivalent to the average price which prevailed in the mid-1970s in real terms. This is significantly less than the 1980-81 peak when prices were over $50 per barrel in constant 1990 terms, but higher than the pre-1973 level of about $10 per barrel. Clearly, the future path of crude oil prices is likely to be anything but smooth and will depend on a variety of factors including the rate at which crude production capacity is expanded.

A second important assumption underlying the IEA outlook relates to economic growth. Economic activity in the reference case is assumed to expand at an average rate of about 2.7 per cent per year for the OECD, 3.1 per cent for the USSR and eastern Europe and 4.6 per cent for the developing countries over the period from 1989 to 2005.

It is also worth noting that the projections relate only to the supply and demand of commercial fuels. Non-commercial fuels, which are used extensively in some developing countries, are excluded from the analysis. The projections assume, however, that incremental energy demand in developing countries will be met increasingly by commercial energy substitutes.

Finally, as with any forecasting exercise of this nature, much caution should be employed when interpreting individual numbers. The outlook which follows is intended to be only indicative of the general direction and possible evolution of worldwide energy trends and the more disaggregated the numbers, the less significance should be attached to their individual magnitude.

TOTAL NON-OECD

The non-OECD "region", representing as it does some 85 per cent of the world's sovereign states, is marked by enormous diversity of economic and political systems and levels of economic development. These include the low and medium income economies of Africa, Latin America and parts of the Asia-Pacific region, the dynamic Asian economies (DAEs) undergoing rapid industrialisation, the high income Middle East oil exporters and the emerging market economies of the USSR and eastern Europe. China, which accounts for a substantial proportion of non-OECD population and energy consumption, presents an additional set of economic parameters. For purposes of examining the interplay between the OECD group of countries and the rest of the world it is useful to consider the non-OECD as a single integrated region but in order to examine the dynamics of the non-OECD the Secretariat groups these countries into geographic regions exhibiting similar characteristics. The following outlook assumes that economic activity will expand by 3.1 per cent in the USSR and the economies of eastern Europe and by 4.6 per cent in the developing countries, including China, over the period to the year 2005. This compares with a forecast growth rate of 2.7 per cent for OECD countries.

Many factors will influence the future level and pattern of energy consumption and production in the non-OECD world, each of which will be more or less applicable to any individual country or regional grouping of countries. Key among these factors will be the following:

(i) *economic growth:* while the levels of economic growth assumed in the Secretariat's outlook will necessarily have a direct impact on the level of energy consumption, this will be subject to any changes in a variety of factors, including shifts in industrial structures, the efficiency of the capital stock, such as automobiles and industrial processes, and energy pricing policies. As well as the direct impact of growth on energy consumption, the growth process and its accompanying increases In per capita income will normally also entail a series of sub-processes which will influence both the level and pattern of energy demand. Prime among these will be increased demand for transportation, both for personal use and the movement of goods. The share of transport in total final energy consumption is likely to increase over the forecast period and the bulk of future incremental demand in liquid fuels is likely to stem from growth in the transportation of goods and people. The extent of the impact of increased demand for transport on fuel consumption will be modified to the extent that improvements in vehicle efficiency filter down from OECD countries to the non-OECD region. Of similar importance will be the trend towards urbanisation evident in most developing countries as industrial expansion provides employment opportunities in urban agglomerations. Commercial energy consumption is traditionally higher in urban than rural households, partly as a result of an income effect and also because accessibility to electricity and electrical appliances is greater than in non-urban areas. Urbanisation and rising incomes also contribute traditionally to a shift away from the consumption of non-commercial fuels, including wood and other biomass, towards commercial substitutes such as kerosene and gas. This upward pressure on commercial fuel demand will be strongest in Africa, China and parts of the Asia-Pacific region where non-commercial fuels can account for a substantial proportion of the energy consumed in non-urban households.

(ii) *population growth rates:* while population increases in the USSR and eastern Europe are expected to be in line with those of the OECD (0.5 to 0.7 per cent per year), those in the developing regions of the world are assumed to be much higher, from 1.2 per cent per year in China to 2.9 per cent in Africa and 3.0 per cent in the Middle East. The latter will be more the result of the inflow of expatriate workers than of indigenous increases in the population. Although the impact of population growth on energy consumption will be influenced by any improvements in energy intensity, the sheer growth in the numbers of global energy consumers will place heavy pressure on energy resources.

(iii) *capital scarcity:* while all the non-OECD regions, with the possible exception of parts of the Asia-Pacific area, have extensive natural energy resources, many countries, at least outside the Middle East oil exporters, face major financial constraints on their development. This is especially so in Africa and parts of Latin America where the overhanging debt burden and often poor economic prospects increase the difficulties of raising private investment funds. In the state-controlled or transitional economies of China, the USSR and eastern Europe, energy investments have been subject to cutbacks as governments have implemented more restrictive macro-economic policies. In addition, the prospects for foreign investment in many of the non-OECD economies are limited by institutional constraints such as rules on taxation and repatriation of profits and by inefficient administered pricing systems for many energy commodities.

(iv) *environmental constraints:* while environmental considerations are likely to constrain consumption of fossil fuels to a greater degree initially in the OECD world, the same constraints will eventually operate to a degree in certain non-member areas. Energy-related environmental degradation has been given little attention to date in most non-OECD countries but increasing international concern regarding the global effects of energy consumption may come to have greater bearing on non-member country policy-making over the forecast period. Initially, however, it is likely that local environmental concerns, such as urban pollution from vehicle exhausts and the inefficient burning of dirty fuels in households and industries, will be of greater concern than global issues such as climate change. Siting issues are also likely to become more important, particularly in relation to large hydroelectricity schemes and nuclear power plants. One of the major areas of concern in some non-member countries such as China and many of the African states will remain land degradation and desertification as a result of fuel-wood consumption. Depending on the extent and speed of policy responses to these problems, it can probably be expected that there will be some substitution of more environmentally benign fuels compared with the current situation. This is likely to be complicated, however, by the possible migration of "dirty", energy-intensive types of industrial activities from the OECD area as these economies become increasingly environmentally restrictive at home.

Total Primary Energy

Energy demand has grown rapidly in the non-OECD world in recent years, almost doubling its absolute level in the 16-year period from

1973 to 1989. This rate of growth easily outpaced that of the OECD and, as a consequence, the share of these countries in global energy demand increased to 49 per cent in 1989. The Secretariat's energy outlook suggests that non-OECD primary energy demand will grow at 3.4 per cent per year through the year 2005, to reach 6544 Mtoe, compared with 3827 Mtoe in 1989. While this growth in demand will be less than that experienced in the preceding two decades, it is expected to be considerably higher than the OECD's forecast rate of growth of 1.3 per cent per year. As a consequence, the non-OECD is expected to account for more than half of global energy consumption by the mid-1990s and for about 57 per cent in 2005.

Figure 18: **SHARES OF WORLD TPES, 1989-2005**

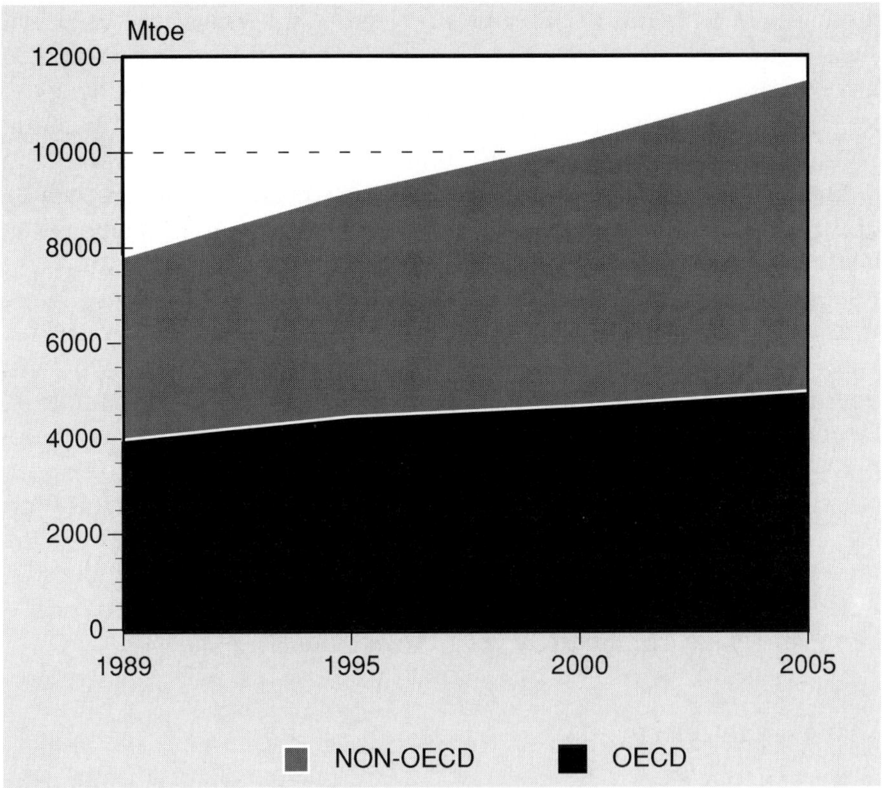

Source: IEA Secretariat

Over the forecast period, the non-OECD fuel demand mix is expected to shift towards the consumption of natural gas, which will account for 29 per cent of total energy consumption in 2005, compared with 23 per cent in 1989. This will reinforce the trend already evident in the 1980s. Oil's share of energy consumption is likely to decline from 36 per cent to 33 per cent over the period, while that of coal is expected to fall from 36 per cent to about 32 per cent. Hydro and nuclear power will account for around 6 per cent of energy consumption throughout the projection period. While oil will remain one of the principal elements of global energy demand, its dominance will vary significantly between the world's regions. Over the period to 2005, it will account for the major share of fuel consumption in the Middle East, Latin America and Africa. Coal will remain the principal fuel source in China and eastern Europe,

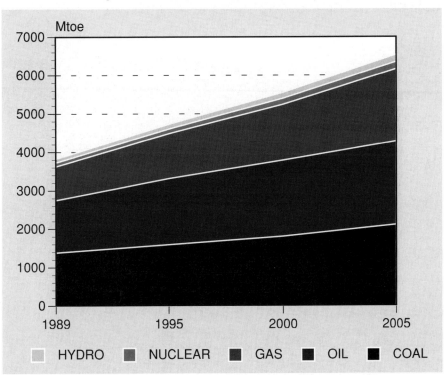

Figure 19: **NON-OECD TPES BY FUEL, 1989-2005**

Source: IEA Secretariat

reflecting natural endowments, and will become dominant in the Asia-Pacific region by 2005. In the USSR, the leading role is expected to be played by natural gas.

Geographic shifts in the share of non-OECD energy demand are also likely to occur. The USSR is expected to maintain its dominance, although its share of the non-OECD total will decline from 36 per cent in 1989 to 31 per cent in 2005. The already small share of eastern European demand is also expected to fall, while the major gains will be made by the Middle East and Asia-Pacific. When the two largest

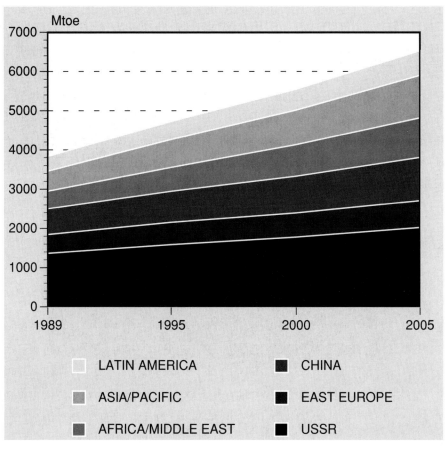

Figure 20: **NON-OECD TPES BY REGION, 1989-2005**

Source: IEA Secretariat

consumers of energy, the USSR and China, are considered together, it is clear that they will continue to have an overwhelming influence on the non-OECD and, indeed, the global energy balance.

Total primary energy production in the non-OECD countries is expected to grow 3.1 per cent per year during the projection period, reaching 8144 Mtoe in 2005, compared with 5005 Mtoe in 1989. Among the individual regions, the Middle East is expected to exhibit the fastest rate of growth at 4.6 per cent per year and will be responsible for almost one third of the incremental increase in energy output. The non-OECD is expected to increase its already substantial share of global primary energy production to 65 per cent and its exportable energy surplus is expected to reach 1600 Mtoe by the year 2005, compared with 1179 Mtoe in 1989. As a consequence,

Figure 21: **SHARES OF WORLD ENERGY PRODUCTION, 1989-2005**

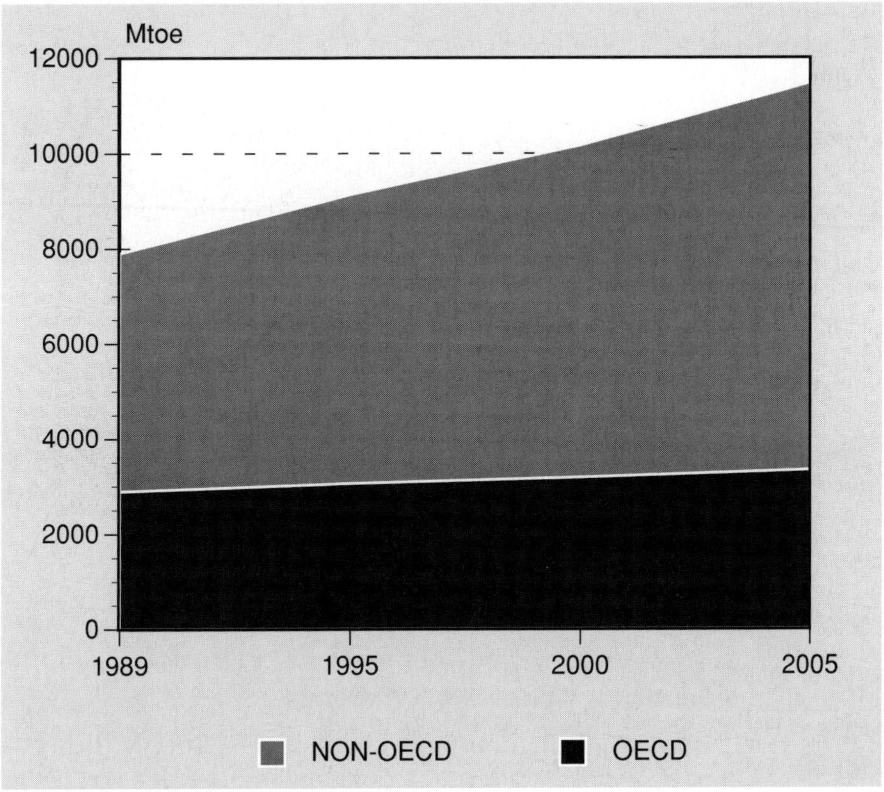

Source: IEA Secretariat

the proportion of OECD primary energy supply met by imports from non-OECD countries is likely to increase to about 32 per cent from 28 per cent in 1989.

In terms of its fuel structure, non-OECD primary energy production is also likely to shift towards natural gas, which is expected to account for 26 per cent of total production by 2005, compared with 20 per cent in 1989. The shares of coal and oil are expected to decline, although oil will remain the dominant fuel, accounting for 44 per cent of output in 2005 and coal for 26 per cent. The shares of nuclear and hydropower in total energy production are expected to remain small.

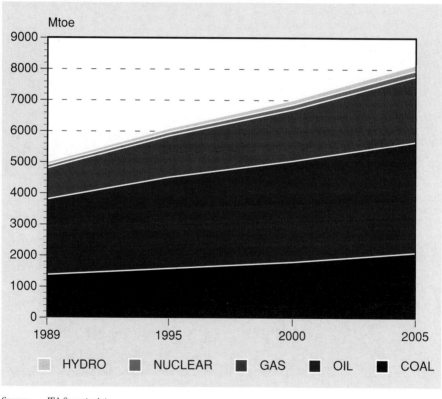

Figure 22: **NON-OECD ENERGY PRODUCTION BY FUEL**

Source: IEA Secretariat

Energy intensity in OECD countries is expected to continue to decline at an annual rate of about 1.4 per cent, reinforcing the trend evident over the 1980s. This drop is due to continued technological advances, structural transformation of OECD economies towards less energy-intensive sectors and continued investments in energy efficiency. It also reflects the long-term price effects and replacement of old capital stock with newer, more energy efficient equipment. In the non-OECD countries, by contrast, energy intensity is expected to remain consistently higher over the forecast period. While the reasons for this situation will vary from one region to another, they will include such factors as the development or persistence of energy-intensive industrial structures, outdated energy transformation and consumption technologies, inefficient or subsidised energy price mechanisms, and pressure from population growth. The Secretariat's outlook assumes that energy intensities in Latin America, Africa and probably the Middle East will increase. The ongoing political and economic developments in eastern Europe and the Soviet Union make it possible that measures to improve energy efficiency may be introduced, particularly in the area of energy pricing. In China, with very high but declining intensity, energy conservation is expected to remain one of the pillars of national energy policy and it is likely that intensity will continue to decline. In the Asia-Pacific region, where levels of economic development vary widely between countries, energy intensity is expected to decline, although more moderately than in the OECD region, as a result mainly of transformations in industrial structures.

Regional Overview

While general trends in the non-OECD are of interest, it is obvious that the characteristics of each region, including its level of development and national income, its industrial structure and resource base, will have a significant impact on the development of its energy balance. While the individual regions are treated in detail in later sections of this chapter, some of the major factors are highlighted below.

In the Middle East, energy demand is expected to grow more strongly than elsewhere in the non-OECD region, at an annual rate of 6.9 per cent. This growth will be underpinned by population increases, a rapidly expanding industrial sector, particularly in energy-intensive refining and petrochemicals, and increased demand for transportation. Attention will be placed on increasing the use of natural gas in the domestic market, in order to free as much oil as possible for export, and the region is expected to increase its dominance of the world oil market. By the year 2005, up to 50 per cent of non-OECD oil production, or 38 per cent of global production, could be generated in the Middle East.

The Asia-Pacific region will also experience large increases in energy consumption, based on continued high rates of economic growth. The region is expected to remain extremely dependent on the rest of the world for its energy supplies, with net imports quadrupling over the projection period. This will be partly a result of the expected peaking of indigenous oil production. Securing energy supplies to fuel the economy will remain a particularly important policy objective and, coupled with increasing environmental concerns, the region is expected to make efforts to diversify its energy economy.

Latin America's economic and energy outlook will be influenced overwhelmingly by its attempts to reduce the region's huge accumulated external debt, which was responsible for a considerable slowdown in economic growth in the 1980s. Political and social stability will also be important factors. It is anticipated that a gradual recovery in economic performance will occur over the projection period and that energy demand will grow moderately at 3.3 per cent per year. While the bulk of world oil production is expected to be concentrated increasingly in the Middle East Gulf region, Venezuela is also likely to retain its position as an important supplier of oil to world markets.

In Africa, energy production and consumption is dominated by a small number of countries and this situation is unlikely to change. Despite the large energy surpluses that the region as a whole is expected to generate, many sub-Saharan states will continue to experience energy shortages and non-commercial fuels will form an important part of their energy balances. In all, energy demand in Africa is projected to grow 3.0 per cent per year. The region's exportable energy surplus, however, is likely to increase considerably, second in volume terms only to that of the Middle East.

The Soviet Union is currently the world's largest producer and second largest consumer of energy. The ongoing political and economic transformation process is imposing enormous economic dislocation and the period at least to the mid-1990s will be a time of substantial transition. The energy economy will inevitably be affected by these processes but a healthier economy and energy sector may emerge in the post-transition period. Indeed, if the current political and economic reforms are successful and lead to improved relations with the western world, including increased economic co-operation and foreign investment in the development of resources, the Soviet Union could become, in the long term (perhaps beyond the outlook period of this report), a substantial supplier of energy to world energy markets and an important alternative to the Middle East. The uncertainty surrounding these possibilities, however, remains high and prospects for the immediate future indicate some further declines in both production and consumption of energy.

The outlook in eastern Europe will be influenced by considerations similar to those operating in the Soviet Union, although individual circumstances will obviously vary by country. Assuming a successful transition to more market-oriented economic systems, it is expected that energy demand will continue to grow. Energy production, however, may stagnate over the projection period as a result of investment constraints, resulting in a sharp increase in energy imports.

The main impetus to energy demand growth in China over the outlook period will be the expected high level of economic growth. Securing adequate fuel supplies to underpin this growth will remain an important objective of national energy policy. Energy demand in the world's most populous nation is projected to grow 3.4 per cent per year. While oil exports are an important source of hard currency on which China relies to import foreign expertise and services as well as goods, the exportable surplus is expected to be squeezed by increased domestic demand. The country is likely, however, to remain a small net exporter in the year 2005.

An Overview by Fuel

(i) Coal

Coal consumption in the non-OECD region as a whole grew more slowly than other fuels between 1973 and 1989 and is likely to remain moderate over the forecast period. Growth is expected, however, to be higher than in the OECD countries and, hence, the non-OECD's share of global coal consumption will rise to about 63 per cent in 2005. Among the non-OECD regions, the developing countries, including China, are expected to almost double their consumption of coal between 1989 and 2005, which will represent an actual deceleration in the growth rate compared with the previous decade. Expansion is expected to be most vigorous in China and the Asia-Pacific region, where the power generation sector will remain the overwhelmingly dominant consumer.

Coal production in the non-OECD area is expected to grow at an annual rate of 2.6 per cent over the projection period, with output increasing from 1385 Mtoe in 1989 to 2110 Mtoe in 2005. The non-OECD's share of global coal production will rise from 60 per cent to 63 per cent. In eastern Europe and the USSR, coal production as well as consumption is likely to stagnate through the mid-1990s. The concentration of coal burning facilities in locations with declining production has produced strains on an already overburdened coal transportation system and, before expansion of coal production can resume, an improvement in both the location of industry and in the transportation infrastructure will be required. It is not expected that coal output in these regions will rise significantly until the early years of the next century. China, with the world's largest hard coal reserves, is expected to continue its vigorous expansion of production and South Africa is expected to account for an increasing share of incremental output gains.

Coal reserves are more widely dispersed geographically than other fuel types, which, coupled with vastly differing coal qualities and requirements, has resulted in extensive international trade. The non-OECD region as a whole has been a small net exporter of coal and this situation is likely to continue. A substantial increase in the exportable surplus is expected to come from Africa, specifically South Africa, while Latin America is expected to become a small net exporter during the 1990s as a result of production increases in Colombia and Venezuela. The resource-poor Asia-Pacific region is expected to increase its net imports quite rapidly in order to meet increasing demand from the industrial and power generation sectors.

(ii) Oil

World oil consumption increased 15 per cent during the period 1973 to 1989, all of which was accounted for by increases in non-OECD countries. Oil demand in the non-OECD region is expected to remain strong over the projection period, growing at an average annual rate of 2.9 per cent. Consumption is expected to increase to 2173 Mtoe in 2005 compared with 1365 Mtoe in 1989. Consumption growth in the OECD area is expected to be only 0.6 per cent per year and, as a consequence, the share of the non-OECD in global oil consumption will rise from 43 per cent in 1989 to about 52 per cent in 2005. Within the non-OECD, the developing countries, particularly China, the Asia-Pacific region and the Middle East, are expected to display the highest rates of growth. The rise in oil demand is expected to come mainly from the transport sector and the petrochemical industry, while the use of heavy fuel oils in power generation is expected to decline. As a result, there will be a substantial shift in non-OECD demand towards the lighter end of the oil barrel which may require substantial upgrading of existing refining capacities, as well as the construction of new capacity. The consumption of gasoline and diesel fuel, for example, is expected to increase by about 75 per cent in these countries over the projection period.

The non-OECD region accounts for 94 per cent of world proven oil reserves which are estimated at about 1000 billion barrels[1]. Two-thirds of the world total and 70 per cent of the non-OECD total lie within the Middle East region.

Oil production in the non-OECD is projected to grow 2.4 per cent per year, to reach 3560 Mtoe in 2005, compared with 2429 Mtoe in 1989, an absolute increase of 47 per cent. This compares with an expected decline in oil output of 23 per cent over the same period in the OECD region. As a result, the non-OECD's share of world oil production is expected to rise from 76 per cent in 1989 to 86 per cent in 2005. Within the non-OECD region, however, most of the increase will

---

1. All data pertaining to oil and natural gas reserves are from *Oil and Gas Journal* (31 December 1990). All reserves figures except those for the USSR are reported as proven reserves recoverable with present technology and prices. USSR figures are "explored reserves" which include proven, probable and some possible reserves. Reserves/production ratios are from *BP Statistical Review of World Energy*, June 1991, based on *Oil and Gas Journal* data.

come from the Middle East, with Latin America contributing a minor proportion. Indeed, the Middle East region alone is expected to account for 38 per cent of world oil production in 2005 compared with 26 per cent in 1989. Production in the Soviet Union is likely to continue its decline in the short term before resuming growth in the late 1990s and output from eastern Europe is expected to fall gradually throughout the projection period. In the Asia-Pacific region, which currently accounts for only 6 per cent of the non-OECD total, production is expected to peak by the turn of the century. Ultimately, the level of non-Middle East oil production will depend on reserve additions, which in turn will depend not only on the evolution of crude oil prices but also on technological advances in exploration, development and production techniques, and on other circumstances, including government policies.

Overall, the non-OECD's exportable oil surplus is expected to increase substantially, reaching 1387 Mtoe by 2005, compared with its 1989 level of 1064 Mtoe. At the same time, the OECD countries are expected to become increasingly dependent on imports of oil and it is possible that about 70 per cent of OECD consumption will be met from imports from non-OECD countries. This compares with a figure of 58 per cent in 1989. It is inevitable, given the evolving pattern of international oil supply, that this will entail increasing reliance on Middle East oil production.

(iii) Natural Gas

Increases in natural gas consumption in both the USSR and eastern Europe as well as in the developing countries have constituted the most significant change in the non-OECD energy balance over the period 1973 to 1989. Demand for gas almost tripled in this period and its share of TPES increased from 15 per cent to 23 per cent. This increasing dominance is expected to continue in the period to 2005, with demand for gas increasing at an annual rate of 4.9 per cent to reach 29 per cent of TPES in that year. In absolute terms, consumption is expected to more than double from 878 Mtoe in 1989 to 1881 Mtoe in 2005. This compares with a lower rate of increase in the OECD region of 2.3 per cent per year. As a result, the non-OECD's share of world natural gas consumption is expected to reach 63 per cent by 2005, compared with 53 per cent in 1989.

Within the non-OECD region, demand for natural gas is expected to be strongest in the developing countries, with consumption almost tripling over the projection period. The major limiting factor in developing the large gas reserves in these countries is the pace of investment in gas-consuming equipment, although funds are being allocated increasingly for this purpose in some areas. About 45 per cent of the expansion in consumption in developing countries is expected to come from the Middle East, with Latin America and Asia-Pacific each accounting for around 21 per cent. In the USSR and eastern Europe growth is expected to be more moderate.

The increased use of natural gas is expected to be underpinned by demand from the power generation sector, where gas will increasingly replace oil, and by industry, particularly the petrochemical sector. In the power generation sector, consumption is expected to more than quadruple in developing countries and more than double in the USSR and eastern Europe.

The increased penetration of natural gas can be attributed, in part, to policies in oil-producing countries, especially the Middle East, to free as much oil as possible for the export market. Environmental considerations will also become more important over the forecast period in the non-OECD world and are likely to provide a stimulus for increased use of more environmentally benign gas in all sectors.

The production of natural gas in the non-OECD area is also expected to increase rapidly, at 4.8 per cent per year, from 992 Mtoe in 1989 to 2110 Mtoe in 2005. This compares with a more modest annual rate of increase in the OECD region of 1.9 per cent. As a result, the non-OECD region could account for as much as 71 per cent of world gas production by 2005, compared with 60 per cent in 1989.

Among the non-OECD regions, the developing countries are expected to more than double their output over the period to 2005, while production in the USSR is expected to rise about 90 per cent in absolute terms and that in eastern Europe about 8 per cent. Given its large base, the increase in the USSR, although lower than the developing countries in percentage terms, is expected to account for about half of the absolute increase in non-OECD production.

World proven natural gas reserves are estimated at about 4208 trillion cubic feet, of which 88 per cent lie in the non-OECD area. The Soviet Union alone accounts for 38 per cent of the world total and 43 per cent of non-OECD reserves. As with supplies of oil, the OECD is likely to source an increasing proportion of its natural gas requirements from non-OECD countries. The exportable gas surplus of the non-OECD is expected to double over the projection period, from 114 Mtoe in 1989 to 229 Mtoe in 2005, the largest share of which will be from the USSR.

(iv) Nuclear

The contribution of nuclear power to meeting the energy needs of the non-OECD area increased in recent years, although in global terms nuclear power capacity is still heavily concentrated in OECD countries. The output of nuclear electricity in the non-OECD expanded from 4 Mtoe to 100 Mtoe over the period 1973 to 1989, and is expected to grow at an annual rate of 4.0 per cent over the projection period, to reach 196 Mtoe in 2005. As a result, the non-OECD's share of global nuclear electricity generation is expected to rise from 20 per cent in 1989 to 28 per cent in 2005.

Safety concerns in the post-Chernobyl era and the fact that nuclear energy no longer enjoys undisputed cost advantages over its alternatives for the generation of baseload electricity have dampened the prospects for more rapid increases in nuclear output. Despite major cancellations and lengthening of already considerable construction times, the USSR and eastern Europe are expected to account for about half the net additions to world nuclear output and about 90 per cent of those in the non-OECD. The dominance of this area in nuclear power developments to 2005 seems assured because of the sheer mass and scope of their current construction programmes but the current political and economic changes have clearly increased the uncertainty of the projections for these regions.

The prospects for nuclear power in developing countries are more conservative and it is expected that the contribution of nuclear power to total energy supply in these countries will remain insignificant. The expansion of this capital-intensive energy form with its technological complexity and need for highly trained personnel implies a high dependence on external sources and a drain on scarce hard currency earnings which only a small number of developing countries can seriously contemplate.

(v) Hydropower

The production of hydropower has grown strongly in the non-OECD region, more than doubling between 1973 and 1989. The Secretariat's outlook indicates that output will grow at 4.1 per cent per year, from 98 Mtoe in 1989 to 185 Mtoe in 2005. This growth will account for about 75 per cent of the global increase in production. Accordingly, the non-OECD's share of the total is expected to rise from 47 per cent in 1989 to 55 per cent in 2005.

Eastern Europe and the European section of the USSR are characterised by a mature stage in the development of hydroelectric power and, hence, most of the non-OECD increase is likely to occur in the developing countries, including China, where the untapped hydro potential is still large. One factor acting to constrain the rapid and extensive expansion of hydro capacity is the fact that unrealised potential will become increasingly concentrated in more remote areas where development is more difficult and expensive. Financial constraints will also be significant in many countries.

(vi) Electricity

Demand for electricity is expected to remain strong in the non-OECD region, reflecting the ongoing processes of industrialisation and urbanisation. In many developing countries, electricity supply is limited to the more densely populated urban areas and there is enormous potential to expand the electricity grid. Rural electrification programmes receive high priority in many national energy policies. Electricity consumption in the non-OECD region as a whole is expected to grow 4.5 per cent per year, more than doubling over the projection period, from 4734 TWh in 1989 to 9370 TWh in 2005. This compares with a 50 per cent rise in the OECD. By the year 2005, about half of global electricity production is expected to be consumed in the non-OECD region, compared with 42 per cent in 1989.

Table 3: **Non-OECD[a] — Primary Energy Balance[b]**
Mtoe

	1989	1995	2000	2005
Coal				
Production	1385	1561	1780	2093
Net Imports	-1	32	26	17
Consumption	1385	1593	1806	2110
Oil				
Production	2429	2930	3247	3560
Net Imports	-1064	-1207	-1265	-1387
Consumption	1365	1723	1982	2173
Gas				
Production	992	1333	1651	2110
Net Imports	-114	-164	-198	-229
Consumption	878	1169	1453	1881
Nuclear	100	126	165	196
Hydro	98	123	151	185
Total				
Production	5005	6073	6994	8144
Net Imports	1179	-1340	-1437	-1600
Consumption	3827	4733	5557	6544

a Includes eastern Germany
b Excludes non-commercial fuels

Note: Because of rounding, totals and sub-totals may not exactly equal the sums of their individual components

Source: IEA Secretariat

AFRICA

From an economic as well as an energy perspective, the African continent can be described as one of diversity. While the Secretariat considers the region as one unit for general forecasting purposes, it is useful when analysing energy trends to group the countries into three main sub-regions:

(a) the major oil exporters, including Algeria, Egypt, Nigeria and Libya;

(b) a major coal exporter, South Africa; and

(c) the rest of sub-Saharan Africa, including about 40 oil importing countries with generally severe economic problems, including declining per capita incomes and high and increasing debt to GDP ratios.

Notwithstanding the diversity in economic and energy structures in the region, the Secretariat projects an average annual increase in GNP over the period from 1989 to 2005 of 2.9 per cent per year. This compares with expected annual population increases of 2.9 per cent, implying no increases in real per capita incomes.

Total Primary Energy

TPES in the region is expected to increase 3.0 per cent per year — slightly higher than the GNP growth rate — from 220 Mtoe to 354 Mtoe. The region as a whole is clearly dominated by South Africa and a small number of oil exporters. South Africa alone accounted for about 44 per cent of TPES in 1989 and the four major exporters for an additional 38 per cent. When the other exporters are included, the share of total regional consumption rises to 86 per cent, leaving the many sub-Saharan states responsible for only 14 per cent of energy demand, despite the fact that their share of population is about 60 per cent.

The fuel structure of the region's TPES is expected to be increasingly skewed towards natural gas, accounting for about 21 per cent of total consumption in 2005, compared with 15 per cent in 1989. The gains in the share of gas in energy consumption are expected to be made largely at the expense of oil, whose share is projected to decline from 43 per cent in 1989 to 39 per cent in 2005. The share of coal is also

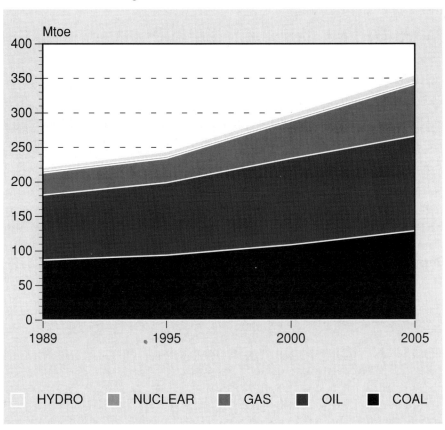

Figure 23: **AFRICA — TPES, 1989-2005**

Source: IEA Secretariat

expected to decline over the period, from 40 per cent to 36 per cent, and neither hydropower nor nuclear power is expected to increase its low proportions of total energy consumption.

Total primary energy production is projected to rise 3.3 per cent per year, reaching almost 800 Mtoe by 2005, compared with 473 Mtoe in 1989. Oil is likely to remain the major fuel but its dominance will decrease. Africa as a whole is a significant energy exporter — in 1989 its net exports of oil were more than double its domestic consumption level and represented 13 per cent of internationally traded oil. Net exports of coal accounted for 10 per cent of world

Figure 24: **AFRICA — ENERGY PRODUCTION, 1989-2005**

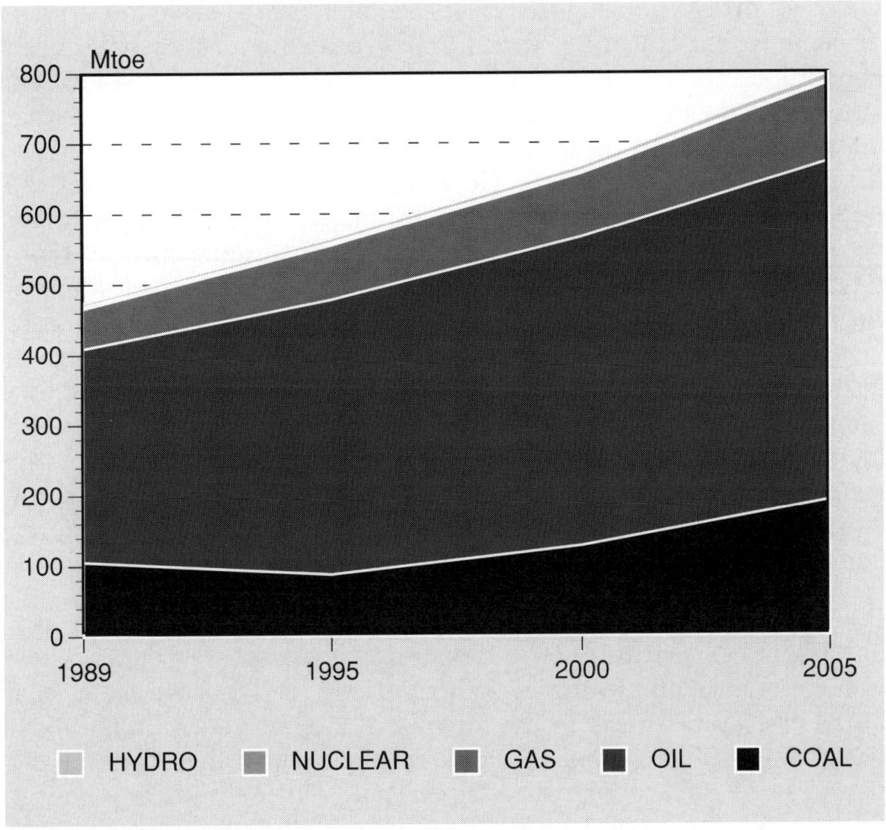

Source: IEA Secretariat

coal exports and those of gas for 11 per cent of internationally traded gas. Overall, the region's exportable energy surplus is expected to rise considerably, to 443 Mtoe from 254 Mtoe in 1989.

The African region as a whole is well-endowed with natural energy resources, although these are unevenly spread and many remain undeveloped. A number of important factors will bear on the ability of the region to develop its energy resources and on the likelihood of these reaching the international marketplace. These include:

(i) *general economic conditions and the investment climate:* the economic performance of the majority of African nations has gradually deteriorated over about the last two decades. Following a period of growth in the 1960s, the 1970s were characterised by stagnation, and in the 1980s per capita GDP actually declined in many countries. Lacking the ability to finance large-scale investment projects from domestic sources, most African countries have been constrained to rely upon foreign capital sources for the exploration and development of their energy resources. The result has been that many countries are now heavily indebted, severely limiting their future borrowing capacity. However, at least 50 per cent of the region's countries have embarked on major structural adjustment reform programmes as a condition of continued support from international lending organisations. The early success of some of these programmes may indicate an increasing ability to generate investment funds from domestic sources as well as an improvement in the climate for foreign investors. The latter will also depend to a large extent on the willingness of governments to relax the institutional constraints on foreign investment, including rules on the repatriation of profits and taxation structures, and to reform the existing inefficient administered pricing system for many energy commodities.

(ii) *population pressures:* one of the most important factors impinging on African energy demand in the coming years will be population growth. In the period from 1965 to 1980, average annual population growth was in the vicinity of 2.7 per cent. Unlike most other developing regions, Africa's rate of growth increased in the period from 1980 to 1988, to 3.2 per cent. The Secretariat's energy outlook for the period to 2005 assumes a slowing in the rate to 2.7 per cent per year. While both African total and per capita energy intensities are low and are likely to continue to be so, the sheer growth in numbers of energy consumers will place heavy pressure on energy resources and, at the margin, will act to limit the availability of energy exports.

(iii) *declining availability of fuel wood:* not unrelated to the population issue is the question of the availability of non-commercial fuel wood as an energy source. The Secretariat's forecasts are for commercial energy only but in many countries non-commercial fuel wood supplies a substantial proportion of primary energy. For the continent as a whole, non-commercial fuels were equal to 33 per cent of TPES in 1989 but in individual countries, such as Benin and Tanzania, the proportion exceeded 85 per cent. The rate of growth of fuel wood supply is currently less than increases in demand and is leading to depletion of available resources. The combination of this situation with increasing urbanisation is likely to lead to a shift in demand into commercial substitutes such as kerosene and gas.

Coal

In 1989, over 90 per cent of African coal consumption and more than 95 per cent of coal production was attributable to South Africa. It is also the world's third largest coal exporter, after Australia and the United States. The only other coal producer of any significance in the region is Zimbabwe, with 3.0 per cent of the regional total in 1989. Botswana and Swaziland also contain some coal reserves. Limited domestic demand, and lack of capital, trained labour and infrastructure, as well as high transport costs will continue to restrain development in these countries. The outlook for coal in the region will continue to be dominated by the outlook for South Africa.

The Secretariat's projections point to a steady increase in coal production over the period, at an average annual rate of 3.8 per cent. This would lead to a rise in output from 106 Mtoe in 1989 to 192 Mtoe in 2005, or 6 per cent of expected world coal production in that year. Net exports are also expected to be substantial. There is no physical constraint on the sustainability of this forecast as South Africa is estimated to contain some of the largest coal reserves in the world. In addition, production costs are low because of thick, easily accessible coal seams and relatively low labour costs. Efficient coal transportation systems are also in place. The major constraint on South African coal production in the past has been unfavourable political circumstances which have limited export markets. The current easing of this situation is likely to lead to increased South African export and, hence, production opportunities.

Oil

The output of oil in the region is expected to grow at 2.9 per cent per year over the forecast period, from 303 Mtoe in 1989 to 481 Mtoe in 2005. Consumption growth is expected to average 2.5 per cent per year and to reach 129 Mtoe in 2005. It is anticipated that the exportable surplus will rise from 210 Mtoe to 344 Mtoe over the same period.

As a result of increasing costs and declining returns on investment due to lower oil prices over the last decade, Africa's importance in the world oil industry has declined. A shrinking proportion of world exploration activity has been devoted to Africa which has resulted in stagnation in the continent's reserves. These represented 6 per cent of proven world oil reserves at the beginning of 1991, compared with 9 per cent in 1980. The share of African production has also declined, from 10 per cent of the world total in 1980 to 7 per cent in 1989.

Nearly 90 per cent of Africa's proven oil reserves of 60 billion bbls are found in the four major producing countries. Of these reserves, 23 billion are in Libya, 17 in Nigeria, 9 in Algeria and 5 in Egypt. These countries are relatively well placed geographically to export oil across the Atlantic to North American markets, or, with the exception of Nigeria, via the Mediterranean to southern European markets and further north by pipeline. Algerian, Libyan and Nigerian crude oil is mainly light and low in sulphur.

Recent petroleum exploration activities have been more successful in Egypt than elsewhere in the group. The Egyptian oil industry is state owned but joint ventures with both Arab and international partners have increasingly been encouraged. The government has created relatively attractive operating conditions in order to maintain a high level of investment in the sector. One of the main tenets underlying Egyptian energy policy is the substitution of gas for oil in domestic consumption in order to conserve crude oil for export and to prolong the life of reserves. Although production levels have probably peaked, a combination of substitution and more rational domestic fuel pricing policies could permit Egypt to increase its oil exports.

Algeria's oil reserves are far more limited than its gas reserves (reserves/production ratios of 22 and 69 years, respectively) and, hence, it has had little incentive to expand oil production. Given the very high dependence of the economy on the hydrocarbon sector, the fall in oil prices in 1986 had a deleterious effect on general economic conditions. In these circumstances, the government undertook wide ranging economic reforms, including the liberalisation of the upstream oil sector and the easing of foreign investment restrictions. The results of the subsequent increase in exploration activity are expected over the next few years.

Exploration activity in Nigeria has declined substantially since the early 1980s as a result of falling oil prices, and output has fallen from its peak in 1979. Nigeria's stated intention is to increase production capacity from the current 1.8 mbd to 2.5 mbd and to raise proven reserves to 20 billion barrels. The potential exists for this expansion to be realised but only with substantial foreign investment. The National Nigerian Petroleum Company (NNPC) has insufficient financial resources to fund expansion on its own account and it is anticipated that it will be required to release more equity in oil projects to fund the necessary investment. A precedent was recently established when NNPC released 25 per cent of its 80 per cent stake in an upstream venture accounting for half the country's total production. Confidence in the Nigerian investment environment is indicated by the range of foreign companies with significant exploration and development plans.

Of the smaller producers, Angola in particular is expected to boost output on the basis of new offshore developments. Despite adverse political conditions, the Angolan government has taken steps to ensure that the country remains attractive to foreign investors and is opening up new offshore acreage to exploration for the first time. It is likely that reserve additions will continue to be made.

In the existing large oil producers, development costs and risks are sufficiently low to encourage further exploration activity and it is likely that this will be focused on known offshore fields, particularly in the Gulf of Guinea. However it is also considered in the industry that important oil reserves remain to be discovered in some of the relatively unexplored areas outside the Gulf. In interior countries, transport difficulties across frontiers to local markets or to coasts for export have also been a disincentive to exploration. Exploration and development of oil and gas resources in Africa has traditionally been dependent on foreign participation to provide funds and expertise. Continued or increased foreign activity, particularly in new, relatively unexplored areas, will require the establishment of favourable investment environments, with particular importance attached to taxation and royalty policies. In addition, many potential oil basins extend over several countries and co-operation among countries — including consistent legislation and arrangements for regional transport corridors — would facilitate exploration in these areas.

Natural Gas

Natural gas is expected to play an increasingly important role in the African energy balance over the forecast period, increasing its share of TPES from 15 per cent in 1989 to 21 per cent by 2005. Total consumption is expected to grow 5.5 per cent per year to reach 75 Mtoe in that year. Production is also expected to rise but at a slower rate than demand. In 1989, net exports of gas were 25 Mtoe, equal to 78 per cent of domestic consumption, and in 2005 are expected to be 37 Mtoe, or about half the domestic consumption level.

Africa's total proven reserves of natural gas are estimated at 285 trillion cubic feet, equal to about 7 per cent of proven world reserves at the beginning of 1991. The distribution of these reserves across the continent is uneven, with almost 70 per cent found in Algeria and Nigeria alone. A further 20 per cent of reserves are in Libya and Egypt. The rate of exploratory drilling activity in Africa has lagged behind those in other regions in recent years and large areas remain unsurveyed. It is possible that the reserves estimates could be revised significantly upwards if hydrocarbon exploration activity intensifies.

Prospects for the continued expansion of natural gas in both domestic and export markets are positive in the large gas-producing countries. In Algeria, for example, official estimates point to a tripling of domestic gas consumption by 2000, particularly in electricity generation and the chemical industry. Algeria is also Africa's largest user of natural gas in the residential sector and compressed natural gas was introduced to the market as a motor fuel in 1989.

In 1989, Algeria exported about 60 per cent of its gas production and it is the export market which will provide the major impetus to expansion of the industry. Exports take place both as LNG and by pipeline to the Italian mainland through Tunisia and Sicily. Most LNG exports are to western Europe, with France being by far the largest purchaser. Other important LNG markets are Belgium, Spain and the United States. The first shipments to Japan were made in 1989. The government has ambitious plans to raise LNG exports over the next several years and the expansion of LNG capacity is currently being studied. The upgrading and demothballing of existing capacity is already being undertaken. The expansion of the gas pipeline system is also being examined, including a proposal for a Transmaghrebine gas system linking Algeria, Tunisia and Libya. A 2000 kilometre pipeline has also been proposed from Algeria to Spain via Morocco, to be constructed as a joint venture between the Algerian and Moroccan governments. Plans also exist to expand the pipeline capacity between Algeria and Sicily and the Italian mainland.

In Egypt, current energy policies and development plans call for greatly increased gas production and consumption in coming years. Incentives for gas exploration were introduced in 1987 which could result in substantial increases in reserves by 2000. Current Egyptian policy does not envisage the export of natural gas until substantial additional reserves have been discovered. In the interim, gas will be substituted increasingly for oil in domestic consumption, freeing oil for the export market. A large number of gas development schemes are currently under way, including the development of six new fields, which are expected to quadruple production capacity in two years. In addition, in November 1988 the government invited bids for six natural gas and oil concessions offshore in an area where substantial volumes of gas have been discovered. Foreign participation will remain an important element of gas sector development.

Nigeria has substantial reserves of natural gas, about half of which are associated with oil deposits. In 1988, only 21 per cent of gross gas production was marketed due to an extremely high level of gas flaring. In an effort to reduce the level of flaring, the NNPC in 1988 invited all companies to submit proposals for the utilisation of associated gas. These are currently under review. The government is also understood to be considering a range of incentives to make gas production investments more attractive. A comprehensive national gas policy is also being considered by government to encourage exploration and gas development activities.

The potential for increased domestic consumption of gas in Nigeria is considerable, with various projects currently under way. In particular, the National Power Authority's Igbin power station near Lagos, Nigeria's largest, has been supplied with gas since late 1988 via a 355 kilometre pipeline. Almost 60 per cent of Nigeria's electricity is now gas generated and the guaranteed gas supplies by the pipeline permit the uninterrupted supply of electricity to Lagos for the first time. While the National Power Authority is the pipeline's primary customer, the NPCC is confident of securing a range of industrial customers along the pipeline route, including the expanding petrochemical industry and the cement, aluminium and glass industries. A substantial LNG export project is also being developed between the NPPC and a consortium of foreign partners. Construction of a liquefaction plant was scheduled to begin in 1991 and shipments are expected to commence in 1995.

Nuclear

South Africa is the only country within the African region with nuclear generating capacity. In 1989, nuclear accounted for less than 3 per cent of South African primary energy supply and 1 per cent of the total region. It provided 7 per cent of South Africa's electricity generation. Despite its position as a major uranium producer there are no current plans to increase nuclear output over the forecast period and no known plans for other countries to commence production. The share of nuclear power in the region's energy supply is, therefore, expected to decline.

Hydropower

Hydropower accounted for only 2 per cent of African energy supply in 1989 and about 18 per cent of total electricity generation. This latter figure was down from 32 per cent in 1980 as a result of increasing access to thermal power sources and a constraint on funds for the development of large-scale hydropower projects. In some countries, however, particularly in the sub-Saharan region, hydropower contributed over 90 per cent of total electricity generation.

Almost two-thirds of the hydroelectricity produced in 1989 was generated in five countries — Egypt, Ghana, Mozambique, Zaire and Zambia. Hydropower output is expected to increase at an average annual rate of 4.9 per cent over the forecast period, from 5 Mtoe in 1989 to 11 Mtoe in 2005, and its share of TPES is expected to rise to 3 per cent. In both Zambia and Zimbabwe, significant undeveloped potential exists on the Zambezi river, jointly shared by the two countries. In Kenya, plans call for the addition of 140 MW hydro capacity by 1995, equal to almost one-quarter of the existing installed capacity. Further development of the hydro potential of the main river basin — the Tana — totalling 500 MW and another 400 MW elsewhere in the country is also under consideration. In Egypt, hydro potential is already highly developed along the Nile basin and most incremental electricity demand is met by conventional thermal sources. There are, however, proposals to expand hydro generation by the installation of low-head systems on existing dams.

The major constraint to the expansion of hydropower generation in the African region will continue to be financial. While small-scale hydro projects can be a cost-effective means of bringing electricity to isolated communities, electrification programmes are often low on the expenditure priorities of the poorer African countries.

Electricity

Electricity consumption in Africa is expected to grow faster than TPES, at an average annual rate of 3.8 per cent over the outlook period. Over half of the region's electricity generation in 1989 was coal-fired and about 49 per cent of regional output was from South Africa. South Africa's dominance is unlikely to diminish in the years ahead and the majority share of coal in regional electricity production is also unlikely to change. There is a possibility elsewhere in the region, however, that natural gas will be substituted increasingly for oil in power generation in order to free oil for export markets. The share of gas is expected to rise to 20 per cent in 2005 compared with 11 per cent in 1989, while that of oil is expected to fall to only 5 per cent, compared with 16 per cent in 1989.

Table 4: **Africa — Primary Energy Balance**[a]
Mtoe

	1989	1995	2000	2005
Coal				
Production	106	88	128	192
Net Imports	-19	6	-19	-63
Consumption	87	94	109	129
Oil				
Production	303	390	438	481
Net Imports	-210	-285	-313	-344
Consumption	94	105	125	137
Gas				
Production	57	77	90	111
Net Imports	-25	-41	-36	-37
Consumption	32	35	54	75
Nuclear	3	3	3	3
Hydro	5	7	8	11
Total				
Production	473	564	667	797
Net Imports	-254	-320	-368	-443
Consumption	220	244	299	354

a   Excludes non-commercial fuels

Note: Because of rounding, totals and sub-totals may not exactly equal the sums of their individual components

Source: IEA Secretariat

ASIA-PACIFIC

The Asia-Pacific region is one of the least homogeneous of the broad non-OECD groups, incorporating countries at widely different levels of economic development and with different social and political systems. The fastest growing economies in the region over the past two decades have been the so-called dynamic Asian economies (DAEs) — Hong Kong, Malaysia, Singapore, South Korea, Taiwan and Thailand. Despite some slowdown from peak rates in the mid- to late 1980s, it is expected that these countries will continue to enjoy high rates of economic growth. It also seems likely that the remaining countries of the Association of South East Asian Nations (ASEAN), namely Brunei, Indonesia and the Philippines, will grow strongly. Industrialisation in Indonesia and the Philippines has been relatively slow to date but it is expected that both countries will experience relatively high economic growth, based on the development of a labour-intensive industrial sector. The outlook for India, with the largest population in the region, is for lower although steady economic growth. The Indian sub-continent was affected by the Gulf crisis through the loss of workers' remittances but lower oil prices and reconstruction in the Gulf are expected to bring relief. Throughout the period to 2005, most countries in the region will experience increased levels of urbanisation, industrialisation, demand for private transportation and rural electrification, all of which will lead to increases in total and per capita energy demand. It is also important that the export orientation of many of the industrialising Asian economies leaves their level of economic growth particularly vulnerable to developments in the world economy, particularly in their major trading partners, the United States and Japan. For the purposes of its outlook the Secretariat has assumed that economic output in the region as a whole will increase at an average annual rate of 5.0 per cent until 2005.

Total Primary Energy

Energy demand in the Asia-Pacific region has grown more strongly than in any other non-OECD region except the Middle East, averaging about 6.0 per cent per year since the early 1970s. Given the high levels of economic growth which are anticipated over the outlook period, this trend is expected to continue. The Secretariat assumes that energy demand in the region will grow at a rate of 4.9 per cent per year to 2005. In absolute terms, TPES is expected to more than double, from 509 Mtoe in 1989 to 1086 Mtoe in 2005. As a result, the region's share of world total primary energy supply is expected to increase from 7 per cent to 9 per cent over the same period.

In comparative terms, energy demand in the non-OECD Asia-Pacific was about equal to that of Japan in the mid-1980s. Reflecting its high relative growth rates, the Secretariat's projections indicate that, sometime during the period 2000-2005, it will probably be twice that of Japan. India alone accounted for 34 per cent of regional energy consumption, as well as production, in 1989 and this dominance is likely to persist into the next century. Energy consumption in the DAEs, which accounted for about half the regional total in 1989, is also expected to continue growing strongly.

Oil and coal will continue to dominate the region's energy demand structure, accounting for about 80 per cent of TPES throughout the projection period. Coal's share of TPES is likely to increase, however, at the expense of oil. By 2005, coal is expected to account for 41 per cent of TPES and oil for 39 per cent. The contribution of natural gas to TPES is also expected to rise, from 9 per cent in 1989 to 14 per cent in 2005. The share of nuclear power is likely to fall to about 3 per cent by the end of the outlook period and that of hydropower is expected to remain constant at around 4 per cent.

Total primary energy production is expected to rise more slowly than consumption, averaging 3.8 per cent per year. In absolute terms, production is expected to reach 783 Mtoe in 2005, compared with 432 Mtoe in 1989. The bulk of this expansion is likely to come from coal and natural gas. Crude oil production in the region is expected to peak around the turn of the century and to stagnate thereafter.

Figure 25: **ASIA-PACIFIC — TPES, 1989-2005**

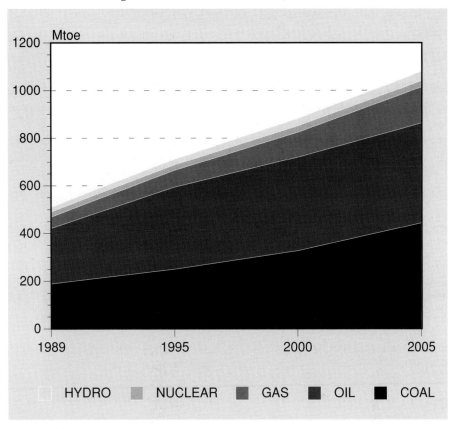

Source: IEA Secretariat

The Asia-Pacific region is a net importer of energy, and the projected differential growth rates between production and consumption will accentuate this position. In absolute terms net imports are expected to increase from 77 Mtoe in 1989 to about 300 Mtoe in 2005. Under this scenario, import dependence will rise from 15 per cent of total energy demand in 1989 to 28 per cent by 2005. As a result, it is likely that security of supply will remain a key concern in the energy policies of many countries. Increased efficiency of energy use will undoubtedly play an important role in developing sustainable energy strategies and increased conservation measures are expected to be introduced in many countries. Structural change towards less energy-intensive industrial sectors will also play a role in the more mature economies. As a result of these trends, the Secretariat's outlook assumes that energy intensity in the region as a whole will decline by an average of 0.2 per cent per year between 1989 and 2005.

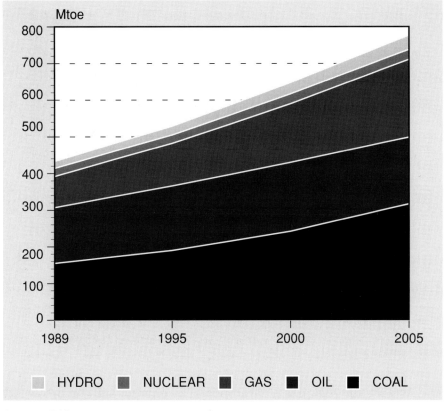

Figure 26: **ASIA-PACIFIC — ENERGY PRODUCTION, 1989-2005**

Source: IEA Secretariat

Coal

Regional coal demand has grown strongly, particularly during the 1980s when large-scale increases in coal inputs to the power generation sector were made in many countries in line with fuel switching policies away from oil. Growth averaged 6.4 per cent per year between 1980 and 1989 and is expected to continue at a rate of 5.5 per cent per year through 2005. On this basis, absolute demand will more than double, from 189 Mtoe in 1989 to 444 Mtoe by 2005. It is projected that the absolute demand for coal will increase more than for other fuels, and that coal will account for 43 per cent of the increase in energy demand during the projection period. As a result, coal's share of TPES will rise to 41 per cent in 2005, compared with 37 per cent in 1989.

The most substantial increase in coal consumption is expected to be made in the power generation sector. India, for example, the largest producer and consumer of coal in the region, is likely to triple its coal inputs to electricity generation by the turn of the century. South Korea, one of the world's largest coal importers, intends to quadruple the use of coal for power generation by the end of the decade, with all incremental consumption likely to be met by imports. Indonesia, currently neither a large producer nor consumer of coal, intends to expand both its production and consumption of coal. Reserves are large and will be used to replace declining indigenous oil supplies. Thailand also intends to increase its use of coal in the power sector, with the bulk of incremental demand expected to come from imports.

With regard to coal output, indigenous production is projected to grow at 4.6 per cent per year over the outlook period, from 155 Mtoe in 1989 to 317 Mtoe in 2005. There are large reserves of coal in the region which remain to be exploited, particularly in Indonesia and India, but substantial investment in both the production and transport sectors will be required. Indigenous production met about 82 per cent of regional coal demand in 1989, with the remaining imports sourced largely from Australia, North America, South Africa and China. As a result of the differential growth rates between production and consumption, net imports are likely to increase to about 29 per cent of coal demand by 2005.

Oil

Oil demand in the region has remained strong since the early 1970s, growing at 4.9 per cent per year. Since the mid-1980s, growth has reached 8 per cent per year or more in some countries, including South Korea, Taiwan and Thailand. Growth is expected to remain high, with the region providing the main engine for increases in global oil demand. Average annual growth rates of about 3.7 per cent are anticipated in the period to 2005. In absolute terms, demand is expected to increase by 80 per cent over the projection period, from 233 Mtoe in 1989 to 419 Mtoe in 2005, increasing the region's share of world oil demand from 7 per cent to 10 per cent.

The bulk of incremental oil demand is expected to come mainly from the transport sector, reflecting the increasing demand for transport associated with rising per capita incomes. Incremental oil demand in the region is expected to be met, in part, by increases in indigenous production. Oil production is projected to grow at 2.0 per cent per year through the turn of the century but to decline at a rate of 0.6 per cent per year thereafter as reserves are depleted. In absolute terms, this implies an increase in production from 152 Mtoe in 1989 to 188 Mtoe by the end of the century and a decline to 182 Mtoe in 2005. A growing proportion of oil demand in the later period will, therefore, be met by imports. In 1989, the difference between domestic demand and indigenous supply was 81 Mtoe. This is expected to widen over the projection period, to 237 Mtoe by 2005.

Indonesia, India and Malaysia are currently the major oil producers in the region, accounting for about 90 per cent of total regional production. Substantial gains in production have been achieved over the past decade in Malaysia and India, while Indonesia's production has stagnated. The region's proven reserves were 24 billion barrels at the beginning of 1991, equivalent to only 2 per cent of the world total. Over the preceding decade, however, substantial increases in reserves were made only in India. The reserves/production ratio for the region as a whole is estimated currently at about 20 years.

In Indonesia, the majority of existing oil fields have reached a mature stage and, in order to sustain the current level of production, substantial investment in enhanced/secondary recovery will be required. It is generally considered that there remains substantial unrealised potential but recent discoveries have been small. Without major new discoveries, production is likely to decline and the country may find it difficult to generate an exportable surplus beyond the turn of the century. In Malaysia, one of the largest oil producers in the region, production is projected by the national administration to peak by the mid-1990s. Malaysia is expected to become a net importer by 2005. India, with the region's second largest proven reserves, has increased its exploration and development activities under the current five-year plan, which calls for a tripling of drilling activity in that period. There is considerable potential from both existing fields and new discoveries. In addition, there are many areas, particularly offshore fields, which are yet to be exploited. Even if successful, rapidly rising domestic consumption is likely to ensure that India's position as a net importer remains unchanged. Significant increases in output from other countries in the region are unlikely. Combined with increasing regional demand, this situation will inevitably lead to greater reliance on oil imports.

As growth in world petroleum production capacity appears destined to be concentrated ultimately in the Middle East Gulf region, it seems unavoidable that the Asia-Pacific will source an increasing proportion of its oil demand from the Middle East. Already, a number of large oil-consuming countries in the region source over 50 per cent of their imports from the Middle East and in some cases this proportion has been increasing steadily. The maintenance of a secure supply of oil is likely to remain an important policy concern.

A further complicating factor is the likelihood of a shift in the quality of indigenous crude oils towards those which are heavier and of higher sulphur content, as production of lighter and low sulphur content crudes declines in the longer term. This will be compounded by the continuing shift in demand towards middle and light oil products, mainly as a result of increasing demand for transport fuels. Environmental considerations are also likely to shift preferences towards low or unleaded gasoline and low sulphur crudes. These developments are likely to have a marked impact on the refining industry. Refinery capacity in the region is currently very tight, capacity expansion having lagged demand growth over the second half of the 1980s. The growth in transport fuels, in particular, has far outstripped the ability of regional refiners to supply such products to the local market. The loss of Kuwaiti conversion capacity at the beginning of the recent Gulf crisis highlighted the basic tightness of product supplies in the region as those countries highly dependent on Kuwait's product exports sought replacement supplies and raised domestic throughputs where possible.

Given the projections for oil demand and supply in the region, it is clear that there will be need for a substantial increase in distillation and conversion capacity during the coming decade, both in the Asia-Pacific region and in the main region supplying its product imports, namely the Middle East. Within the region, existing projects under construction and firm plans, principally in Indonesia, South Korea, India and Thailand, indicate a probable increase of 0.8-1.0 mbd in regional distillation capacity between 1990 and 1995. This compares with current capacity of 5.1 mbd. Past experience suggests that many other projects under discussion will, in practice, be delayed or cancelled, leading to the conclusion that refining capacity is likely to continue to be tight, with the region becoming increasingly dependent on product imports. In many cases, governments will be required to play a key role in facilitating refinery developments by ensuring that there are no excessive delays in obtaining planning approvals and that commercial terms and environmental requirements are clear.

Natural Gas

The region's demand for natural gas increased rapidly over the 1980s, averaging 14.0 per cent per year between 1980 and 1989. The increase can be attributed largely to fuel diversification policies for energy security reasons and to growing environmental concerns. Within the region, natural gas is most widely used for electricity generation purposes in South Korea, Taiwan and Singapore as well as India, Pakistan and Thailand. It is expected that future growth in demand for gas will be strong, averaging 7.6 per cent per year through 2005. The absolute level of consumption is expected to rise from 47 Mtoe in 1989 to 151 Mtoe by the end of the projection period. As a result, the share of gas in TPES is likely to rise from 9 per cent in 1989 to 14 per cent in 2005.

Natural gas production in the region has shown rapid growth, averaging 15 per cent per year since the early 1970s. Growth is expected to slow to 5.9 per cent per year over the period to 2005 with production rising to 212 Mtoe in that year, compared with 85 Mtoe in 1989. As a result, net exports are expected to rise from 38 Mtoe to 61 Mtoe over the same period.

The region's proven reserves of natural gas are 243 trillion cubic feet, equivalent to about 6 per cent of total world reserves. About two-thirds of these reserves are found in Indonesia, Malaysia and India. The reserves/production ratio is estimated at about 60 years, a much more favourable position than that for crude oil.

With the depletion of its oil resources, especially those of higher quality, Indonesia is encouraging the increased consumption of natural gas for both domestic and export markets. Expanded sales to neighbouring countries as well as Japan are being negotiated. In Malaysia, a fuel diversification strategy aimed at reducing dependence on oil is also expected to boost gas consumption. One of the most ambitious projects currently under construction in the region is a pipeline to connect Malaysia's offshore fields with gas users on the Malaysian Peninsula and, eventually, Singapore. Trade in natural gas is also being negotiated between Indonesia and Singapore, and Thailand and Malaysia. India has stated its intention to more than double gas production and to end flaring by 1995. The power generation and chemical sectors are expected to be the main domestic beneficiaries.

Nuclear

The supply of nuclear energy in the region has grown at an average annual rate of about 24 per cent since the early 1970s, although from a low base. It is projected, however, that the growth rate will decline sharply over the projection period, averaging 2.4 per cent per year through 2005. In absolute terms, demand is expected to reach 33 Mtoe in that year, compared with 22 Mtoe in 1989. Accordingly, the share of nuclear in the region's energy mix is expected to fall to 3.0 per cent by 2005, from 4.0 per cent in 1989.

The only countries in the region with nuclear power capacity are India, Pakistan, South Korea and Taiwan. The Philippines' first nuclear power plant was almost complete in 1986 but construction was halted as a result of post-Chernobyl safety concerns. Indonesia continues to explore the nuclear option. Together with considerable capital costs and the need for technical expertise involved in the development of nuclear energy, an important constraint on the further expansion of facilities in the Asia-Pacific area will be increasing sensitivity to safety and environmental issues. In South Korea and Taiwan, public opposition to the construction of nuclear power plants is now a major factor affecting the pace of capacity expansion. In South Korea, the public movement is strongly opposed to the construction of new plants, while accepting those already in operation. In Taiwan, where public opposition is coupled with cost constraints, the government suspended site preparation for its fourth nuclear plant in 1986.

Hydropower

Demand for hydropower grew 6.7 per cent during the period 1980 to 1989 and it is expected that demand growth will average about 4.6 per cent per year to 2005. In absolute terms, this implies an increase from 19 Mtoe in 1989 to 39 Mtoe in 2005. The share of hydropower in the region's total primary energy supply will remain at around 4 per cent.

The majority of countries in the region have some hydroelectric potential, although its size varies enormously. Estimates of such potential are largest for India, Indonesia and Malaysia and most of the expansion in hydroelectric output is expected to occur in these countries. Most of the potential in Indonesia and Malaysia, however, lies in remote island areas. In Indonesia, for example, about 90 per cent of the hydro potential is located on islands outside Java where the main load centres lie. In Malaysia, about 85 per cent is located in Sarawak and Sabah states where indigenous demand is weak.

While hydropower has to date been seen as a relatively inexpensive source of energy, growing concern over its potential environmental impacts has begun to affect expansion plans in the region, especially of larger projects. The large capital requirements of hydro plants are also a compounding factor. India's hydro potential is the largest in the region but only about 13 per cent has been realised to date. As the second largest source of electricity in the country, the development of hydro resources is expected to continue but the large capital requirements may act as an impediment.

Electricity

With rapid increases in industrial output, as well as increased urbanisation and rural electrification, the demand for electricity in the region is expected to grow strongly. The Secretariat's analysis indicates annual increases of the order of 5.6 per cent per year. In absolute terms this implies an increase from 710 TWh in 1989 to 1694 TWh in 2005. Inputs to conventional thermal electricity generation are expected to shift increasingly away from oil towards coal and natural gas. Coal is expected to account for about 51 per cent of inputs in 2005, compared with 41 per cent in 1989. The share of natural gas is also expected to increase to 12 per cent by the end of the forecast period, compared with 8 per cent in 1989.

Table 5: **Asia-Pacific — Primary Energy Balance**[a]
Mtoe

	1989	1995	2000	2005
Coal				
Production	155	190	242	317
Net Imports	34	60	85	127
Consumption	189	250	327	444
Oil				
Production	152	176	188	182
Net Imports	81	168	206	237
Consumption	233	344	393	419
Gas				
Production	85	116	160	212
Net Imports	-38	-45	-56	-61
Consumption	47	71	103	151
Nuclear	22	28	33	33
Hydro	19	24	31	39
Total				
Production	432	534	654	783
Net Imports	77	183	234	303
Consumption	509	717	887	1086

a Excludes non-commercial fuels

Note: Because of rounding, totals and sub-totals may not exactly equal the sums of their individual components

Source: IEA Secretariat

CHINA

Recent economic growth in China has been among the most rapid of any developing country and has significantly outpaced that of the OECD and Soviet/eastern Europe regions. Increases in economic output averaged about 10 per cent per year over the 1980s, underpinned by a programme of industrial modernisation and reform of the market system. Faced with an overheated economy and rapid inflation in 1988, however, the government introduced a macro-economic stabilisation programme, as a result of which growth has slowed to an annual rate of 4-5 per cent. Growth was also affected by the sharp reduction in foreign investment flows that resulted from the political upheavals in 1989. Stabilisation remains at the forefront of domestic policy concerns but the authorities are moving increasingly to restore economic momentum. The sustainability of acceptable rates of economic growth in the coming decade will hinge on China's commitment to further reform and the openness of its economy. If the reform programme is adhered to, growth of around 6 per cent is seen to be compatible with inflation declining below 10 per cent. Consistent with this position, the Secretariat's outlook assumes that economic activity will grow at an annual average rate of 6.3 per cent during the period to 2005.

The reform programme to date has included the scaling back of central control over the economy by reducing the number of commodities subject to administrative output targets, and the granting of greater freedom in the state enterprise sector to determine the composition and pricing of output, to retain profits and to decide on the disposition of retained earnings. Some product markets have been created which allow producers to trade their above-plan output at freely determined prices and the role of the financial sector has been broadened, reflecting the shift away from complete budgetary support of investment. External trade and exchange rate reforms have had a significant impact and new legislation governing foreign investment has helped attract an increasingly large volume of foreign capital.

As an essential sector, the energy sector has been less subject to reform in the areas of enterprise control and price setting than other areas of the economy. In general, the allocation of products from the energy sector and the provision of investment funds to the sector remain under the control of the central authorities. Lack of co-ordination between sectors has often resulted in bottlenecks in the energy supply system and energy shortages, mainly in the form of electricity blackouts, have frequently resulted in idle industrial capacity and, hence, reduced economic output. Energy prices have generally been kept very low and usually do not reflect costs of production or quality differentials. A limited two-tier pricing system exists in the coal and oil sectors which has boosted output and investment but is also reported to have contributed to the high rates of inflation experienced recently. Overall, prices remain highly distorted and provide little incentive for energy conservation or the rational allocation of energy resources.

Total Primary Energy

Total energy demand in China has been strong in recent years, growing at an average annual rate of 5.8 per cent since the early 1970s. The Secretariat's outlook assumes that growth will decline to 3.4 per cent per year over the period to 2005, accompanied by a substantial improvement in energy intensity. In absolute terms, total primary energy supply is expected to reach 1102 Mtoe in 2005, compared with 649 Mtoe in 1989. China is currently the third largest consumer of energy in the world, consuming more energy than the Asia-Pacific region as a whole. This situation is unlikely to change and the country's share of world total primary energy supply is expected to increase from 8 per cent to 10 per cent between 1989 and 2005.

Although environmental issues are becoming increasingly important, coal is expected to continue to dominate the country's energy mix, accounting for about three-quarters of TPES throughout the projection period. In general, the energy demand structure will remain unchanged, with the exception that nuclear power generation is expected to come on stream for the first time during the early 1990s. Its share of TPES, however, will remain small.

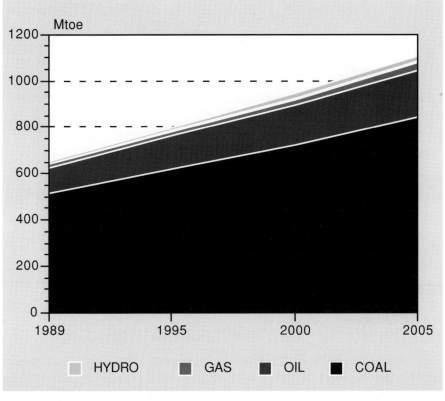

Figure 27: **CHINA — TPES, 1989-2005**

Source: IEA Secretariat

Growth in total primary energy production is expected to average 3.1 per cent per year, to reach 1112 Mtoe in 2005, compared with 679 Mtoe in 1989. Primary energy production will continue to be dominated by coal, representing about three quarters of the total throughout the projection period. China will remain a small net exporter of energy during the projection period but because of the differential growth rates between demand and production, the exportable surplus is expected to decline from 30 Mtoe to about 10 Mtoe.

Figure 28: **CHINA — ENERGY PRODUCTION, 1989-2005**

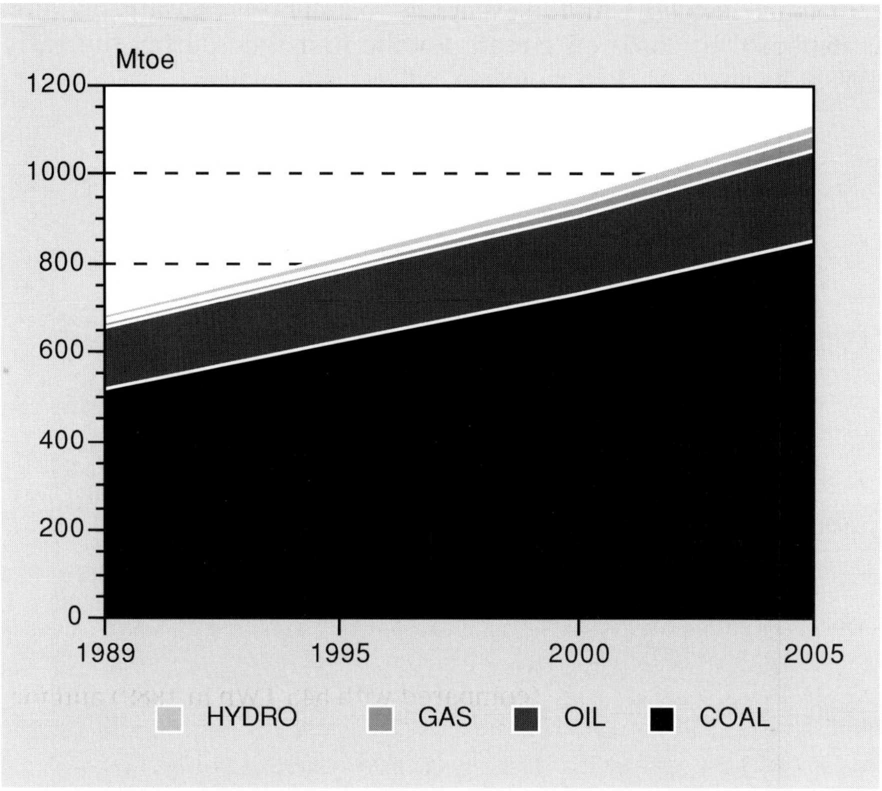

Source: IEA Secretariat

The energy intensity of the Chinese economy has been traditionally among the highest in the world. Since 1980, the government has implemented an energy strategy aimed at reducing waste and encouraging conservation, particularly in the industry sector. In the virtual absence of a market price mechanism this has included the introduction of a range of administrative and legislative measures designed to reduce energy consumption, to encourage more efficient energy technologies, and to implement management and training initiatives. Policy measures to induce a shift in industrial production to less energy-intensive processes and output were also introduced. At the domestic level, large efforts were directed towards the replacement of old and inefficient stoves in both rural and urban households. Substantial progress was made in lowering the economy's energy intensity during the 1980s and intensity declined at a much faster rate than that achieved by any OECD country in a similar period. Over the projection period, energy intensity is expected to continue to decline but at a slower rate.

Coal

Coal consumption in China has grown at an average annual rate of 6.0 per cent since 1973 and the country has emerged as the world's largest consumer of coal, as well as its largest producer. Coal demand is expected to remain strong for the foreseeable future, although its rate of growth is likely to decline compared with the previous period. It is expected to remain the dominant fuel in China's energy balance, accounting for about three-quarters of TPES in 2005. The Secretariat's outlook assumes that coal demand will grow 3.2 per cent annually over the projection period. In absolute terms, consumption is expected to reach 844 Mtoe by 2005, compared with 513 Mtoe in 1989.

Growth in coal consumption is expected to be strong in the power sector. The official government target for electricity generation is 1200 TWh by the year 2000 (compared with 545 TWh in 1989) and for installed capacity to reach 240 GW — an increase of about 100-110 GW from the present situation. Of the newly installed capacity, thermal power is expected to account for more than two-thirds, virtually all of which is to be coal-fired. In order to conserve oil for export, existing oil-fired units are to be converted to coal wherever possible, and, in principle, no new oil-fired plants will be built. The share of electricity generated from coal is expected to exceed 50 per cent in 2005, compared with 41 per cent in 1989. Coal demand will also play an important role in the household sector where coal will substitute increasingly for traditional fuels.

The Secretariat's outlook indicates that the production of coal in China will increase by about 3.2 per cent annually, from 517 Mtoe in 1989 to 850 Mtoe in 2005. It is anticipated that about 50 per cent of production will continue to come from small, locally run mines with the remainder from large commercial or state-run enterprises. According to official estimates, China's proven coal reserves are about 170 billion tons and hence the potential for expansion is significant. An important constraint on the increased utilisation of coal, however, will be the transportation and distribution network. The major share of reserves is concentrated in the north and west of the country, making it necessary to transport large amounts of coal to the main load centres in the east and south. The current transportation system is rail based and is constrained by inadequate infrastructure and rolling stock. Port facilities are also frequently congested which acts as an impediment to efficient export operations. Plans exist to construct additional railroads to the largest coal mines during the 1990s, and coal slurry pipelines are planned in at least two areas. Emphasis is also to be placed on the construction of power stations adjacent to coal mining areas in order to alleviate the pressure on transportation. A further problem in the coal sector is the lack of coal washing facilities which leads to transportation inefficiencies, a reduction in the combustion efficiency of the coal, higher particulate emissions and variable coal export qualities.

Oil

Growth in China's oil demand averaged 4.9 per cent per year between 1973 and 1989 but is expected to be less than this over the period to 2005. The Secretariat's outlook projects that demand will grow at an average annual rate of 3.6 per cent, from 114 Mtoe in 1989 to 201 Mtoe in 2005. As a proportion of total energy consumption, oil is expected to remain constant at around 18 per cent. As noted above, the consumption of oil in the power generation sector is officially discouraged but it is expected that its use in industry, particularly the petrochemical sector, will increase. The petrochemical industry is currently expanding rapidly and official plans indicate a substantial increase in petrochemical production during the 1990s. Oil consumption in the transport sector is also likely to grow, with some shift away from the bicycle towards the motorcycle as a mode of private transport.

In order to meet the expected shifts in oil product demand, a considerable expansion and upgrading of the domestic refinery system will be required. The system is currently geared mainly to the throughput of heavy, low sulphur domestic crudes for the production of fuel oil. As demand for gasoline and diesel fuel increases, as well as demand for petrochemical feedstocks, the refinery system is likely to come under increasing pressure. While China has traditionally been a net exporter of oil products, it is possible that this situation may be reversed if the capacity and flexibility of the refining system is not enhanced.

On the output side, production of oil is expected to increase by 2.4 per cent per year through 2005, to reach 205 Mtoe in that year compared with 140 Mtoe in 1989. This is very close to the government's target for oil production of 200 million tons by the year 2000.

While its ultimate recoverable reserves are still to be defined by comprehensive geological surveys, China's proven oil reserves at the beginning of 1991 were estimated to be 24 billion barrels, sufficient for about 23 years' consumption at current levels. Oil fields in the northeast and eastern regions of the country are estimated to contain about 85 per cent of total reserves and currently provide about 90 per cent of production. Fields in the western region are also considered to have significant potential. It is anticipated that about three-quarters of the increases in output will continue to be sourced from the eastern fields, which are relatively close to major consuming centres, with the remainder from the western region.

For China's oil production targets to be met, continued exploration and development will be essential. Distribution infrastructure requirements, particularly for pipeline transport, are large and will require extensive construction work. With increasing cuts in the government's investment budget, the possibility of expanding domestic financing is limited. Low administered prices for oil are insufficient to provide the upstream sector with investment funds. While foreign investment in offshore areas has been encouraged since the early 1980s, it has not made a significant contribution to oil production and the government's current attitude to foreign investment remains to be clarified.

China is currently a net oil exporter and oil has been an important source of foreign currency earnings. In view of the increasing domestic demand for petroleum products, however, the volume of crude oil and products available for export is likely to be squeezed. Overall, the Secretariat expects China's exportable oil surplus to decline from 26 Mtoe in 1989 to about 4 Mtoe by 2005.

Natural Gas

Demand for natural gas in China has grown strongly since the early 1970s, averaging 5.8 per cent per year. Its contribution to total energy consumption, however, has remained small — in 1989 natural gas accounted for only 2 per cent of total primary energy supply. Given the government's stated policy intention to increase the role of gas in domestic energy consumption, it is expected that growth will remain strong, at 6.4 per cent per year to 2005. On this basis, gas will supply about 3 per cent of total energy needs in 2005.

Growth in gas consumption is expected to be strongest in the industry sector, particularly in petrochemicals. Official estimates, for example, indicate that by the end of 1995 the gas industry will be called upon to supply feedstocks for about 20 large chemical fertiliser plants. The expanded use of natural gas in the residential sector is currently constrained by the lack of an adequate pipeline distribution system but this is being addressed slowly and consumption is expected to increase over time. Gas is viewed as a cleaner alternative to the extensive use of coal for domestic energy needs. The consumption of natural gas in the power sector is expected to grow only slowly as emphasis is placed primarily upon coal-fired generating capacity.

China's proven reserves of natural gas at the beginning of 1991 are estimated at 35 trillion cubic feet or about 1 per cent of total world reserves. The Secretariat's outlook assumes that production growth will keep pace with increases in consumption. Most of the country's production is from two major fields — Sichuan Province in the southwest and the Daqing oil field in the northeast. Sichuan is by far the largest gas resource base in China, accounting for about 70 per cent of the country's gas fields. China is currently increasing both its onshore and offshore gas exploration activity in a number of areas in the hope of substituting gas for oil in domestic consumption and thereby increasing its oil export potential.

Nuclear

To date, no nuclear electricity generation capacity has been introduced in China, although two plants are under construction. In view of serious energy shortages, particularly in the eastern region where the economy is comparatively developed, government policy is to develop nuclear power as a means of improving energy supply. The first nuclear power station is expected to be operational before the end of 1992. Although, over the projection period, efforts will be made to expand capacity, the role of nuclear power is expected to remain small. The Secretariat estimates that, by 2005, nuclear output will account for less than 1 per cent of TPES.

Although China's nuclear plans have been scaled back considerably from the early 1980s, the development of nuclear power "bases" is seen as a viable means of solving the serious energy shortages in the major coastal industrial centres. Two projects are currently under construction. The first, at Qinshan near Shanghai, will provide 300 MW of generating capacity and is expected to be operational by the end of 1992. Two further reactors, each of 950 MW, are under construction at Daya Bay in Guangdong Province north of Hong Kong, the first of which is expected to come on line in 1992 with the second unit to follow a year later. Further serious consideration is being given to a second plant in Guangdong of 1200 to 2000 MW capacity and plans for a second phase of the Qinshan project are under consideration.

Hydropower

Consumption of hydropower has exhibited substantial growth since the early 1970s, averaging 7.3 per cent per year. The share of hydropower in China's energy mix, however, has remained small, at 2 per cent in 1989. The Secretariat's outlook assumes that production of hydropower will rise at an average annual rate of 5.6 per cent over the period to 2005, reaching 24 Mtoe in that year. This represents 2.2 per cent of forecast TPES.

China has the largest exploitable hydropower potential in the world, equal to about 380 GW, of which only 8.6 per cent has been exploited to date. About 70 GW of this potential is suitable for the development of small-scale, local power stations, which are expected to play an important role in rural electrification. As well as a number of very large projects, a range of medium scale hydropower stations is expected to be constructed during the 1990s, especially in the energy deficient south, east and northeast regions.

Electricity

The supply of electricity in China remains a serious problem, both for economic development and the quality of life. Currently, for example, about 250 million rural dwellers — one quarter of the population — have no access to electricity on a reliable basis. Electricity shortages have also been a constraint on industrial development and the government places high priority on the expansion of electricity production in its energy strategy. With abundant reserves, emphasis is to be placed on the development of coal-fired and hydropower capacity.

In the Secretariat's scenario, production of electricity is expected to increase at an average annual rate of 6.3 per cent, from 569 TWh in 1989 to 1503 TWh in 2005. Coal will increase its dominance, accounting for three-quarters of total production in 2005, compared with 69 per cent in 1989. The share of hydropower will remain around 18 per cent, while the share of oil will fall from 11 per cent to 5 per cent. As noted, nuclear power generation is expected to commence in the first half of the 1990s although its contribution to total electricity generation is likely to remain minimal.

Table 6: **China — Primary Energy Balance**[a]
Mtoe

	1989	1995	2000	2005
Coal				
Production	517	622	730	850
Net Imports	-3	-4	-5	-6
Consumption	513	618	725	844
Oil				
Production	140	156	175	205
Net Imports	-26	-12	-3	-4
Consumption	114	144	172	201
Gas				
Production	12	16	22	33
Net Imports	0	0	0	0
Consumption	12	16	22	33
Nuclear	0	0	0	0
Hydro	10	14	19	24
Total				
Production	679	808	946	1112
Net Imports	-30	-16	-8	-10
Consumption	649	792	937	1102

a   Excludes non-commercial fuels

Note: Because of rounding, totals and sub-totals may not exactly equal the sums of their individual components

Source: IEA Secretariat

EASTERN EUROPE

Developments since 1989 in the countries of eastern Europe have set in train a complex set of processes including, political, economic and social reforms. Economic reforms to move from a command to a market economy have been initiated with varying speed and depth. The process is most advanced in Hungary, Poland and Czechoslovakia where comprehensive reform programmes have been put in place. These combine macro-economic stabilisation with price liberalisation and the opening of the trade regime to provide the incentives for the necessary changes to the production structure. At the same time, privatisation has been initiated and financial markets have been introduced to enhance responsiveness to these incentives. Although the process of structural transformation will take many years to complete, there is evidence that the programmes in these countries are starting to achieve positive results. While overall economic output has declined — a development considered unavoidable in the early stages of transformation to market economies — there is evidence that the private sector, although still very small, has been growing strongly. Exports from Hungary and Poland have also grown, especially to European Community countries, and there has been a rapid increase in direct foreign investment in Hungary and Czechoslovakia. In Bulgaria and Romania by contrast, the previous command system has largely disintegrated but less has been done to ensure macro-economic stability or to establish the institutional requirements of a market-driven economy. While both countries are committed to reform, the transformation process is likely to take longer than in Hungary, Czechoslovakia or Poland.

The energy sector will be affected by the reforms in many ways and the successful functioning of the sector will be of pivotal importance for the reforms themselves. Prior to the transition process, the energy sectors of the eastern European economies shared a number of characteristics. These include heavy dependence on indigenous lignite reserves and the environmental degradation this has entailed in the absence of effective measures for environmental protection, very high energy intensities, a significant dependence on imports, especially from the Soviet Union, widespread price distortions, and operational inefficiencies. Fundamental to the transition process in all sectors of the economy will be price reforms. In the energy sector, serious price distortions and extensive subsidy programmes have intensified the excessive reliance on low-quality domestic coal, including the continued operation of uneconomic coal mines. They have also encouraged the continued construction and operation of inefficient power plants and distribution networks, as well as inefficient energy-consuming equipment in the industry, transport and residential sectors. As a result, energy intensities have been extremely high — up to twice or three times those of western Europe.

The consequences of the introduction of market-based price systems as well as economic reform and renewal in eastern Europe will be many over the period to 2005. If successful, they will entail increased economic efficiency, higher economic growth, accelerated economic development, higher per capita incomes and lower overall energy intensity. They are also likely to result in the re-orientation of the composition of energy demand towards oil and electricity in order to satisfy increased demand for personal transport and the electrification of households and industry. The consequences of greater pluralistic democracy, increased transparency and freer public debate are also many and include enhanced awareness and concern about the environmental, health and safety aspects associated with the use, particularly, of coal and nuclear power. This may reduce the rate of expansion of these two energy sources from that which had previously been anticipated.

Progress towards a market economy has been made in some of the countries of eastern Europe. However, methodological difficulties abound in establishing a "most likely" case for the medium-term energy outlook, let alone the sensitivity of the energy systems to alternatives around this case. History provides no clear precedents for the transformations which are almost certain to occur in the region. Recognising the uncertainties involved, a cautious approach has been adopted for the purposes of the present outlook which assumes, critically, that the transition to a market system will occur in a generally orderly manner, free from any major disruption. The outlook is based on the assumption that economic output in the region will continue to increase between 1989 and 2005, although growth is expected to be higher in the later part of the period than in the early stages of the reform process.

The present outlook was prepared before the falls in energy production and demand which occurred in 1991 and are expected to continue at least until 1993. GDP growth in the region was also negative in 1990 and 1991 and is expected to be so in 1992. The effect of these recent developments will be to delay any significant increases in energy demand or output, as it is unlikely that 1989 levels of these aggregates will be regained until the mid-1990s. The Secretariat considers, however, that the underlying trends are still valid, although there is likely to be some delay in achieving the forecast levels of energy demand and production. The outlook presented below, however, is not a prediction and has been made on the basis of certain assumptions. The realisation of some or all of these assumptions will depend on future developments.

Total Primary Energy

Given the assumption of rising economic growth and per capita income, the total primary energy supply of the region is projected to increase over the period to 2005. Growth in energy production is forecast to be slower than that of demand, averaging less than 1 per cent per year, more modest than the 1.3 per cent per year realised over the period from 1973 to 1989. As a result of the different growth rates of demand and production, the region's net imports are expected to increase substantially. In 1989, net imports were equal to 123 Mtoe and are forecast to rise to 275 Mtoe by 2005.

The fuel structure of energy demand in the region is expected to remain dominated by coal. Its share in total demand will, however, fall from 57 per cent in 1989 to about 45 per cent in 2005. Most of the gap in demand will be met by oil and natural gas, the latter of which is expected to make a substantial contribution to the absolute growth in energy supply. Nuclear power's share of total energy supply is expected to rise only slightly and that of hydropower will remain insignificant.

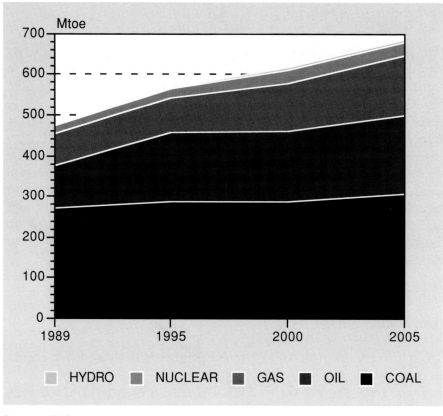

Figure 29: **EASTERN EUROPE — TPES, 1989-2005**

Source: IEA Secretariat

Changes in the production structure are expected to occur to a lesser extent than those in energy supply, with the share of coal remaining at about 75 per cent. The shares of oil and natural gas in total energy production are forecast to fall slightly with some compensating increases in nuclear power output. Coal will continue to make the largest contribution to the absolute increase in energy production.

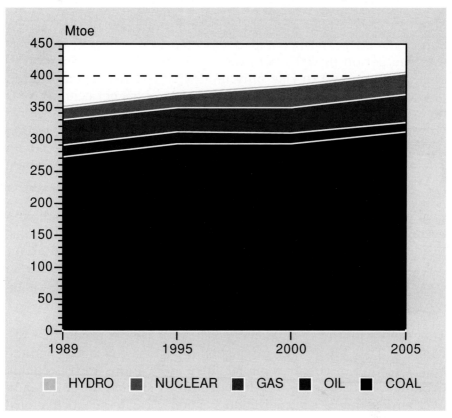

Figure 30: **EASTERN EUROPE — ENERGY PRODUCTION, 1989-2005**

Source: IEA Secretariat

The different rates of growth for economic output and energy demand assumed in the outlook also imply reasonable progress in lowering the region's average energy-intensity. The forces leading to this result are complex, but involve, on the one hand, industrial restructuring away from energy intensive sectors and, above all, price reforms which should induce rationalisation and energy savings. On the other hand, consumer demand for more and better housing and energy-consuming durable goods such as household appliances and motor vehicles is likely to accelerate in the medium term as per capita incomes rise. The net result, according to this scenario, is a decline in overall energy intensity over the period to 2005, with most of the improvement occurring in the second half of the 1990s and beyond.

Coal

Coal has always been the most important element of the eastern European energy balance and is expected to remain so. However, over the period to 2005, both the production and consumption of coal are expected to rise only modestly. Coal consumption is projected to grow at an average annual rate of 0.8 per cent per year between 1989 and 2005, although this disguises slower or negative growth in the earlier part of the period. By the end of the projection period, consumption is expected to reach 308 Mtoe, compared with 270 Mtoe in 1989. As already noted, this implies a fall in coal's share of TPES from 57 per cent to 45 per cent.

Coal production will generally follow the consumption pattern with average annual increases of less than 1 per cent. In absolute terms, production is expected to rise from 273 Mtoe in 1989 to about 312 Mtoe in 2005. The region is expected to remain self-sufficient in coal throughout the period but the level of net exports will remain low.

The reasons for the stagnation of coal use, as well as production, are complex. They involve the exhaustion of mature coal mines and the concentration of coal-burning facilities in locations with declining production. This situation has produced strains on an already overburdened coal transportation system. Despite substantial coal and lignite reserves, it is not clear what proportion of these can be economically mined. Most of the region's high quality coal reserves are located in existing underground mines and are increasingly expensive to extract. In the current reform climate, involving market-oriented pricing regimes and the reduction of extensive subsidies, it is likely that many coal mining operations will be closed.

In addition to economic factors, recent political liberalisation has enhanced public awareness of the severe environmental and health damage caused by the concentration of highly polluting and inefficient coal-burning facilities in the vicinity of population centres. Addressing the environmental and economic problems associated with coal consumption in the region will require restructuring and investment on a massive scale but will constitute a prerequisite for the long-term viability of the coal industry. It should be borne in mind, however, that for most of the countries involved, coal represents a major indigenous energy resource. Given the high level of foreign indebtedness in the region, a role for coal is likely to continue to be sought, albeit at a more rational level.

Oil

Due largely to the heavy reliance on coal and the scarcity of indigenous oil reserves, regional consumption of oil and oil products has historically been low. Consumption is, however, expected to increase over the outlook period. Most of the impetus for growth will come from increased demand for transportation, as well as the impact of the declining rate of growth in coal demand.

Oil production is expected to decline over the outlook period from 18 Mtoe to 15 Mtoe. The region's proven reserves of crude oil are very small, representing about 0.2 per cent of the world total. They are largely concentrated in Romania, Hungary and Yugoslavia and it is only in these three countries that a significant proportion of oil consumption is derived from domestic sources. Output has been declining in Romania for more than ten years and a similar pattern is expected in Hungary. Despite the recent stagnation in Yugoslavian production, some modest increases in output might be realised in the medium term.

Given these circumstances, it is inevitable that the region's dependence on oil imports will increase. In 1989, net imports accounted for 83 per cent of oil consumption and the figure is expected to reach 92 per cent by 2005. To date, the region has been heavily dependent on the Soviet Union as a source of oil imports, supplying about 50 per cent of total imports in 1988. This represented almost 70 per cent of the region's total oil supply in that year. The dependence on the Soviet Union was over 90 per cent for Poland and Bulgaria, and for the land-locked economies of Czechoslovakia and Hungary, the Soviet Union was virtually the sole supply source. Given the current difficulties in the Soviet oil sector, however, exports of crude oil to eastern Europe fell in 1990 and the share of non-Soviet supplies is estimated to have risen to 31 per cent. The principal non-Soviet sources of supply were Iran, Saudi Arabia, Algeria and Libya. Iraq also played a role until the United Nations embargo was implemented late in 1990. The current process of reforming the CMEA trading system is likely to ensure that this shift in trading patterns away from the USSR will continue. The principal reform involves the pricing of Soviet exports in hard currency, a move which was introduced in early 1991. Its main effect is expected to be a substantial rise in the real price of oil imports and a drain on the eastern European economies' limited hard currency reserves. Reflecting this factor, it is expected that barter or other forms of compensation arrangements will continue to be significant in trade between the Soviet Union and eastern Europe. In the longer term, these countries will inevitably seek non-Soviet sources of oil supply. This may have important consequences over time for world oil markets as it will represent an additional source of demand in tightening market conditions.

Natural Gas

As already noted, natural gas is expected to make a contribution to increases in energy demand in eastern Europe over the outlook period. Growth in gas demand is projected to be more moderate than in the recent past as price reform and the introduction of hard currency trade will result in higher natural gas prices to residential consumers, industry and utilities. The increases in gas import prices are expected to be less than those of oil, however, as gas imports from the Soviet Union were priced at less favourable terms than those of oil. In an environment of overall energy price reforms, gas is likely to find increased markets in all the major sectors, but especially in power generation where it is likely to be favoured over coal on environmental grounds. Efforts have also been made in some countries to re-orient household energy consumption towards natural gas.

As is the case with crude oil, eastern Europe has few reserves of natural gas. Proven reserves are currently equal to about 19 trillion cubic feet, or 0.5 per cent of the world total. Gas production is expected to grow only slowly, from 40 Mtoe in 1989 to 43 Mtoe in 2005. As a result, the level of net imports is expected to rise from about half of domestic gas supply in 1989 to over 70 per cent of supply in 2005.

Although gas consumption in eastern Europe will remain heavily dependent on Soviet deliveries for the foreseeable future, some governments in the region have entered into negotiations with other suppliers. Bulgaria, for example, has signed a 20 year protocol with Iran for deliveries and Hungary and Czechoslovakia plan eventual linkages to western European pipelines carrying natural gas from Algeria.

Nuclear

While nuclear output in the region is expected to increase over the outlook period it will be at a significantly lower rate than over the period from 1980 to 1989. Nuclear power currently provides about 11 per cent of the region's total electricity output and this is expected to remain largely unchanged.

One of the most important consequences of increased transparency in decision making and freer public debate in eastern Europe has been to cast doubt on the previously ambitious nuclear programme in these countries and, hence, on the rate of nuclear power development. Nuclear plans had already been considered ambitious before the Chernobyl accident and were considerably scaled down in its wake. But it has only been in the last two to three years, with increased independent inspection and public discussion, that the full scale of the safety risks of some plant designs has been fully acknowledged. Consequently, almost all the programmes are in a state of flux, with major cancellations of planned operations, interruptions to construction, temporary closures and, in some instances, permanent shutdown. In this situation it is increasingly difficult to form a clear picture as to the likely contribution of nuclear power in the region. The Secretariat has built into its forecasts lengthy delays in construction and commissioning which has the effect of pushing capacity increases back towards the end of the present decade. Net additions to nuclear capacity of the order of 40 GW are expected by the end of the outlook horizon.

Hydropower

Owing to unfavourable natural conditions in most countries (the exceptions being Romania and Yugoslavia), as well as to capital shortages, hydroelectricity is unlikely to increase its currently insignificant contribution to the region's energy supplies. The Secretariat's outlook indicates that hydropower production will remain relatively constant over the projection period, its contribution to TPES falling to less than 1 per cent by 2005. The largest projects under construction in the region are located on the Hungarian and Czech side of the Danube river. However, the Hungarian project has been cancelled on environmental grounds and the Czech project is under review. Some potential for small hydro plants has not yet been fully exploited in Poland as well as perhaps in Bulgaria, Romania and Yugoslavia.

Electricity

Since the mid-1970s, electricity production in eastern Europe has grown faster than total energy consumption. This trend is expected to continue. Currently, about 80 per cent of electricity output is derived from conventional thermal power plants, the share of which is projected to rise to about 82 per cent by 2005. A major proportion of incremental output is expected to be generated from natural gas.

Investments of between 5 per cent and 6 per cent of total fixed capital expenditure have been recorded in the electricity sector in eastern Europe in recent years, around the same level as in OECD countries. However, reductions in planned future generating capacity due to the cancellation of several nuclear plants have yet to be made up by planned increases in thermal capacity. Although investments in electricity are now more significant than investment in the coal sector, future power shortages might be expected if the level of economic activity in the region increases significantly or if electricity deliveries from the Soviet Union are reduced. Over the last two decades the Soviet Union has been a net exporter of electricity to eastern Europe, supplying 5.5 per cent of electricity in 1988.

Table 7: **Eastern Europe[a] — Primary Energy Balance[b]**
Mtoe

	1989	2005
Coal		
Production	273	312
Net Imports	-3	-5
Consumption	270	308
Oil		
Production	18	15
Net Imports	86	174
Consumption	104	189
Gas		
Production	40	43
Net Imports	40	106
Consumption	80	149
Nuclear	18	32
Hydro	5	5
Total		
Production	354	408
Net Imports	123	275
Consumption	477	683

a Includes eastern Germany
b Excludes non-commercial fuels

Note: Because of rounding, totals and sub-totals may not exactly equal the sums of their individual components

Source: IEA Secretariat

LATIN AMERICA

The level of economic growth realised in Latin America over the coming 15 years — a critical determinant of total primary energy supply and production — will depend to a large extent on the success of measures taken to reduce the region's external debt as well as on general economic reforms. Virtually all the countries of Latin America are classified by the World Bank as either severely[1] or moderately[2] indebted. While the causes and extent of indebtedness vary from country to country, its impact on the level of growth and investment has been similar across the region. By virtue of the uncertainty and adverse incentive effects it has created, the existence of a large debt overhang has inhibited private investment and the adoption by governments of adjustment programmes.

International developments since 1990, however, under the umbrella of the Brady initiative, point to an increased likelihood of successful debt reduction for the region. Within Latin America, Mexico is the first country to have negotiated a debt service reduction agreement with its commercial creditors which was finalised in 1990 and has already produced encouraging results. The preliminary agreement reached between Mexico and the commercial banks in July 1989 was followed by an immediate reduction in domestic real interest rates as flight capital was repatriated, and by an increase in foreign direct investment. Costa Rica has also finalised an agreement under the Brady plan and Venezuela has initiated debt reduction procedures. Brazil, the most severely indebted economy in the region, will likely follow. In addition, following lengthy periods of sluggish economic

---

1. Defined as countries in which three of four key ratios are above critical levels. These ratios and their critical levels are debt to GNP (50 per cent), debt to exports of goods and all services (275 per cent), accrued debt service to exports (30 per cent) and accrued interest to exports (20 per cent).
2. Debt to GNP (30-50 per cent), debt to exports of goods and services (165-275 per cent), accrued debt service to exports (18-30 per cent) and accrued interest to exports (12-20 per cent).

growth and hyperinflation, several of the region's key economies, including Argentina, Brazil and Mexico, have introduced macro economic stabilisation programmes combined with structural reforms. Positive results have already been observed and are expected to continue into the medium and longer term.

In all, it is assumed that economic activity for the region as a whole will grow at an average annual rate of 3.3 per cent through 2005. With population assumed to rise at an annual rate of about 1.8 per cent, this implies a rise in per capita incomes over the projection period slightly in excess of 1 per cent per year.

Total Primary Energy

Total primary energy supply in Latin America grew strongly throughout the 1970s at a rate of 5.6 per cent per year, and at 2.4 per cent per year over the period from 1980 to 1989. Despite declines in

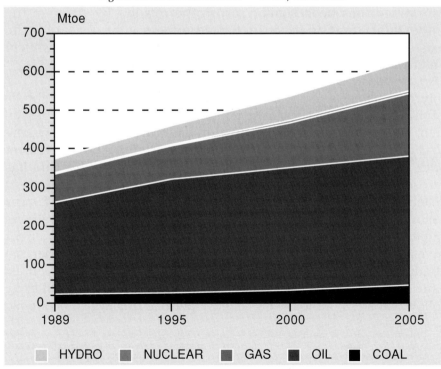

Figure 31: **LATIN AMERICA — TPES, 1989-2005**

Source: IEA Secretariat

its share, the current structure of the region's TPES is heavily dominated by oil, which in 1989 accounted for 63 per cent of energy consumption. Most of the additional growth in supply over the period was met by gas and hydropower, together accounting for 30 per cent of consumption in 1989.

The Secretariat's projections point to an average annual increase in the region's TPES between 1989 and 2005 of 3.3 per cent. In absolute terms TPES is expected to reach 630 Mtoe in 2005, compared with 376 Mtoe in 1989. Since the rate of growth of total energy consumption is above the rate of growth of economic activity, energy intensity is anticipated to increase over this period.

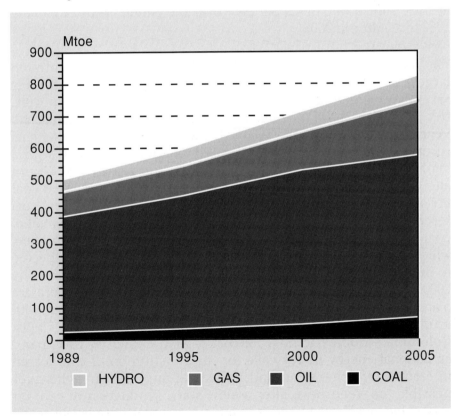

Figure 32: **LATIN AMERICA — ENERGY PRODUCTION, 1989-2005**

Source: IEA Secretariat

Primary energy production is expected to grow 3.2 per cent per year to reach 824 Mtoe in 2005, compared with 500 Mtoe in 1989. Oil is expected to remain the demand fuel in the production structure, although its share is likely to decline. The exportable energy surplus is expected to increase to 194 Mtoe from 125 Mtoe.

Coal

Coal has always played a minor role in Latin America's fuel consumption pattern and this is expected to continue throughout the forecast period. From 6 per cent of TPES in 1989, coal's share of the region's energy consumption is expected to increase to 7 per cent in 2005. The absolute level of consumption will rise at an average annual rate of 4.9 per cent over the period, from 22 Mtoe in 1989 to 47 Mtoe in 2005.

The region's production and consumption of coal is concentrated largely in Colombia, Mexico and Brazil, although Venezuela has recently emerged as a potentially significant producer/exporter. Development plans in the region are mainly centred on Colombia and Venezuela. Elsewhere, the relatively low quality of indigenous coal reserves, their generally inaccessible location, combined with a shortage of efficient transport facilities, and the availability of alternative fuel sources will limit the prospects for increased coal utilisation. Recent developments in Colombia, with the largest coal reserves in Latin America, serve to underline the future potential of coal in that economy. The major project is the El Cerrejon mine, which, together with the El Descanso and La Loma fields, is expected to produce between 15 and 20 million metric tons (mt) per year by 2000. Most of this will be export dedicated. In addition, there are plans to modernise and develop a large number of small mines, with a total production target of 4 mt per year by the early 1990s. Colombia's domestic coal market is also expected to expand as national energy policy calls for the substitution of coal for oil in electricity generation and industrial applications wherever possible. In Venezuela, after a slow start, production of coal for export has accelerated and it is anticipated that output will reach 7 mt by 2000.

Oil

Oil consumption is projected to increase at an average annual rate of 2.1 per cent between 1989 and 2005, the lowest rate of growth of all fuels. This is much smaller than the rate recorded in the 1970s but higher than that of the 1980s. Oil is expected to contribute about 37 per cent of the absolute increase in energy demand over the outlook period, or 94 Mtoe. Despite a decline in its share of TPES, the region will continue to be heavily dependent on oil as its main source of energy.

Indigenous oil production is expected to expand at a slower rate than consumption in the period to 1995 but to outpace consumption growth in the following period to 2005. Production is expected to reach 510 Mtoe by that year. As a result, the exportable surplus is expected to stabilise at about 125 Mtoe in the period to the mid-1990s but to increase to 177 Mtoe by 2005.

With substantial proven oil reserves and large unexplored areas, Latin America has the potential in the medium term to expand its share of international oil supply. At the beginning of 1991 the region's reserves were equal to 121 billion barrels. Its ability to expand supply, however, will depend on the success of national energy policies designed to shift domestic demand away from oil towards gas and hydropower, and, more critically, on the ability and willingness to mobilise both foreign and domestic funds for the large-scale investments which would be required. The severe economic problems and large debt overhang experienced by many of these countries will be a serious obstacle to the mobilisation of investment funds. The fact that, in many countries in the region, the oil industry is largely or exclusively a state-run enterprise, and often closely inter-related with the rest of the economy, leaves investment in the sector vulnerable to restrictive macro-economic policies. In the current climate of fiscal austerity, the availability of government funds for investment may come under pressure. In addition, most oil producers in the region limit, if not exclude, the participation of foreign investment in oil exploration and development activities. The future path of traded oil prices, which will be largely determined by the supply behaviour of the largest Middle East producers, will also act as a critical external constraint on the economic viability of Latin America's resources and on the optimal timing of their exploitation.

The largest oil economies in the region are Venezuela and Mexico, which together accounted for 69 per cent of total regional oil production in 1989. Venezuela's proven oil reserves at the beginning of 1991 were about 59 billion barrels, or about half of those of the region as a whole. By way of comparison, Venezuela's reserves are the sixth largest in the world and in excess of any country outside the Middle East, including the Soviet Union. In addition, the Orinoco heavy oil belt is estimated to contain between 250 billion and 480 billion barrels of recoverable oil, from which Venezuela is currently attempting to commercialise Orimulsion, an emulsion of extra heavy crude oil and water. Only a small proportion of Venezuelan territory has been explored for oil and the potential for further discoveries is thought to be high. At the policy level, the Government has taken rapid steps to allow foreign companies rights to explore for and produce oil and gas, a situation which had not been permitted since the hydrocarbons industry was nationalised in 1976. The state oil company, Petroleos de Venezuela, plans to invest substantially in an expansion of the oil sector, including increases in production capacity, the construction of a new refinery and the expansion of the downstream petrochemicals industry. Foreign participation will be an important component of this programme. A joint venture is also under way to increase the sales of Orimulsion, which are currently very small. Provided that the economic and investment climates remain favourable, Venezuela's potential to increase its share of the international oil market is likely to remain positive. Because of its geographic proximity, however, most increased international activity is likely to be centred on North America.

The prospects for Mexico's future oil output are less favourable. As a result of reduced drilling activity in recent years, production is likely to continue the decline evident since the mid-1980s. While this was partly a result of a short-term policy to restrain output in order to underpin OPEC's price support attempts after 1986, it has had the effect of reducing not only current production but also current and future production capacity. Oil exploration and development activity in Mexico is constitutionally limited to the state oil company (Pemex), to the exclusion of private domestic and foreign investors. The government's current economic reform programme has reduced the availability of investment funds to the oil industry and Pemex has been required to cut its expenditure plans in the medium term.

Thus, while reserves are substantial (52 billion barrels at the beginning of 1991), it is possible that Mexican output could fall severely by 2005. The only foreseeable alternative is that a combination of a slowly changing political environment and some improvement in economic conditions will increase the availability of investment capital, from either domestic or foreign sources.

In Brazil, similar budgetary constraints have operated on the state company, Petrobras, whose rate of drilling fell about 30 per cent between 1986 and 1988 and whose expenditure plans for the post-1990 period have also been cut. Petrobras' financial position is also constrained by government policy to subsidise consumers by selling products in the domestic market at prices which frequently do not cover costs. Sufficient development work has been undertaken in Brazil, however, to ensure that output will continue to rise slowly until the mid-1990s. As well, as much of the Amazon area of the country is relatively unexplored there is potential for large additions to reserves.

In Argentina, which traditionally has attracted higher levels of foreign investment than elsewhere in the region, drilling rates have generally been maintained. The resource base is small, however, and it is unlikely that production capacities and output levels can be maintained beyond the mid-1990s. Prospects for increased oil output in Colombia are also unfavourable, partly because of the limited resource base but also because of the precarious economic situation and the recent tightening of controls over private upstream operations.

Natural Gas

After steadily increasing its share of TPES over the period 1973 to 1989, natural gas is expected to follow the same trend over the forecast period. Consumption of natural gas is projected to increase at an average annual rate of 5.0 per cent, from 75 Mtoe in 1989 to 165 Mtoe in 2005. As such, it is expected to make a substantial contribution to Latin America's incremental energy demand growth. Its share of the region's TPES is expected to increase from 20 per cent in 1989 to 26 per cent in 2005.

With estimated proven reserves of gas equal to 242 trillion cubic feet (about 6 per cent of the world total), there is substantial potential for natural gas to play an increasing role in the Latin American energy balance. The region's reserves are located primarily in Venezuela (106 trillion cubic feet), Mexico (73 trillion cubic feet) and Argentina (27 trillion cubic feet). Production of natural gas in the region is also expected to grow at a rate of 5.0 per cent per year, reaching 166 Mtoe by 2005. Net exports, however, are expected to remain insignificant.

To date, production and consumption of natural gas have been concentrated in the three high-reserve countries, who together accounted for 82 per cent of the region's production of gas in 1989 and 85 per cent of its consumption. Although the levels are much lower, the rate of growth in gas consumption has been higher in the smaller gas economies such as Colombia and Brazil and its contribution to TPES has been increasing at a faster rate in these economies. The energy sector, mainly the oil extraction industry, continues to be the major user of gas in Latin America, reflecting the associated nature of most deposits. The contribution of gas to total electricity generation has remained at about 20 per cent but there has been increased penetration in the industry sector, particularly the chemicals industry, where gas has been substituted for some petroleum-based feedstocks.

While the expansion of gas has been rapid in certain areas of Latin America and the reserve base is sufficient to permit further growth, several factors may continue to act as constraints on its increased penetration. First, the fact that a substantial proportion of gas reserves are associated with crude oil deposits will, to some extent, tie the exploitation of gas to oil production levels. This is especially the case in Argentina and Venezuela. Secondly, the location of many of the region's gas deposits away from major urban and industrial centres limits the immediate potential for increased utilisation unless large investments are made in transportation and distribution networks. Thirdly, and related to the latter, is the capacity of the generally debt-burdened Latin American economies to finance the expansion of gas production and distribution facilities.

Despite these constraints, it is clear that policies in the major Latin American energy economies are directed in part towards increasing the role of gas in their energy structures. In Venezuela, for example, a specific policy to maximise gas consumption has been implemented since 1983. The primary goal is to replace residual fuel oil and diesel fuel in electricity generation and industry in order to free additional petroleum for export without increasing crude production. Significant investments have been made over the period which facilitate the substitution of gas for oil, including the construction of pipelines and connections to electricity stations and consuming areas. The proposed expansion of capacity in the petrochemical industry from 2 mt to 5 mt per year by 1993 is also likely to enhance the future prospects of gas. In addition, Petroleos de Venezuela has recently indicated interest in gaining a share of the United States gas market during the 1990s, including an intention to search for opportunities for integration in the gas sector similar to the downstream joint ventures that it has made with refining companies in host countries.

In Argentina, also, an important element of domestic energy policy is to increase domestic consumption of gas, especially in industry and oil refining, to free up crude production for export. Official national estimates aim for a doubling of both production and consumption of natural gas between 1987 and 2000. To this end the government has implemented a promotional pricing policy to encourage the use of gas in the residential sector and as a substitute for liquid fuels in industry. The government's emphasis on developing a gas pipeline network is a necessary corollary of the gas promotion strategy. As a result of domestic financing constraints, however, major financial and administrative concessions have been offered to attract foreign capital. The government is also concerned to ensure that local companies are involved in joint ventures to develop the country's natural gas potential.

Brazil's strategy towards natural gas is its development for industrial and transport uses, although the latter has been complicated by the government's alcohol fuel programme. Assisted by World Bank finance, a gas distribution project is being developed which seeks to supply gas to industrial, residential and commercial users in the Sao Paulo district, which accommodates 60 per cent of Brazil's industrial output and a population of over 16 million. It also aims to develop efficient gas pricing policies.

In Colombia, where gas has been used predominantly in electricity generation, a wider range of alternative uses is being planned. For example, a pilot project exists in the north coastal region to utilise natural gas for urban public transportation. Colombia is also examining a number of potential industrial uses, including a fertiliser plant and a pipeline from the northern gas fields to the gas-short central regions. Construction is expected to begin on a 700 kilometre pipeline to carry gas to Bogota, which should be fully served by the gas distribution network by 1997.

Nuclear

The nuclear option has been little exploited in Latin America, with Argentina the only producer until 1984, when it was joined by Brazil. Mexico's first nuclear plant, on which construction commenced in 1970, commenced operation in late 1990. Consequently, the share of nuclear in the region's total energy consumption was only 0.5 per cent in 1989. This is expected to rise to 1.3 per cent of TPES by 2005.

Nuclear facilities will continue to be limited for the foreseeable future. In Argentina, two plants are functioning but have been affected by serious operational problems, and a third is under construction. Cutbacks in the nuclear budget since the mid-1980s, however, seem to indicate a change in policy orientation away from nuclear, with increased preference for the exploitation of hydropower potential. Brazil's first nuclear plant reached full commercial operation in 1985 but has been plagued by numerous operational problems since, involving lengthy shutdowns. Construction of a second plant commenced in 1976 and is now about 60 per cent complete, and a third plant was commenced in 1981 on which little work has been undertaken. The future of both these facilities is currently at some risk because of environmental and safety concerns.

Hydropower

Hydropower has traditionally made the third-largest contribution to Latin America's TPES after oil and natural gas, reaching 10.0 per cent of primary energy supply in 1989. This position is expected to continue throughout the forecast period, with hydropower's share of TPES rising to 12.0 per cent in 2005. The absolute level of hydropower consumption is expected to grow at an average annual rate of 4.4 per cent.

In most of the countries of Latin America there is substantial, as yet unrealised hydropower potential. Estimates of total hydropower potential, both realised and unrealised, are largest for Brazil, at 209 GW, followed by 100 GW in Colombia, 80 GW in Venezuela and about 50 GW in Argentina. Energy policies in these countries also incorporate strategies to increase the already significant share of hydroelectricity in total electricity output. In Venezuela, for example, loan finance from the Inter American Development Bank is used to further develop the country's hydroelectric potential. Since 1984, a "small" hydropower programme has been established, the intention of which is to supply electricity to remote areas in the Andean region. Target areas are those with no electricity supply and those with old, thermal plants with capacity of less than 10 kW. In Brazil, the current electricity supply plan to 2000 envisages the expansion of two major hydropower plants, including the Itaipu station — a joint investment with Paraguay — which is the largest in the world. When completed, these two plants alone are expected to provide capacity of over 20 GW. It should be noted, however, that the initial capital costs of hydropower plants are very high and that a large part of Brazil's foreign debt is due to heavy investment in hydro schemes. In Colombia, where more than two-thirds of the installed electricity generating capacity is hydro-powered, there are plans to build several new hydroelectric plants by 1995, with a combined additional capacity of 5.6 GW. At the same time, plans call for only an additional 800 MW of thermal generating capacity which would increase hydro's share of total capacity to about 80 per cent. The state electricity authority, however, has substantial foreign debt and projected operating deficits which may make it difficult to finance a rapid expansion of its hydroelectric facilities. Argentina's situation is similar, with two major hydro projects under construction but which have been hampered by financial difficulties. Together, these facilities would provide additional generating capacity of about 5 GW. In most countries of Latin America where major hydropower schemes are proposed there is evidence of increasing environmental opposition.

Electricity

Electricity production is expected to grow 5.2 per cent per year in this region, from 612 TWh to 1370 TWh during the projection period. As noted above, electricity production in the region has been characterised by the dominance of hydropower which provided about two-thirds of total electricity output in 1989. This dominance is likely to persist in the years to come, but, with the increasing use of natural gas, hydro's share is expected to fall to about 57 per cent.

Table 8: **Latin America — Primary Energy Balance**[a]
Mtoe

	1989	1995	2000	2005
Coal				
Production	22	32	46	64
Net Imports	0	-7	-11	-17
Consumption	22	26	34	47
Oil				
Production	365	418	480	510
Net Imports	-126	-123	-161	-177
Consumption	239	295	319	333
Gas				
Production	75	90	117	166
Net Imports	1	-2	-3	0
Consumption	75	88	114	165
Nuclear	2	5	8	8
Hydro	38	48	60	76
Total				
Production	500	593	711	824
Net Imports	-125	-131	-175	-194
Consumption	376	462	535	630

a   Excludes non-commercial fuels

Note: Because of rounding, totals and sub-totals may not exactly equal the sums of their individual components

Source: IEA Secretariat

MIDDLE EAST

The Middle East region will enjoy relatively strong economic growth in the 1990s, based on increased demand for oil and gas exports, development and diversification of an energy-intensive industrial sector and the mobilisation of private sector resources. The availability of cheap and abundant oil and natural gas reserves has, in the past, led to the development of a large, energy-intensive industrial sector in the region. As a means of decreasing their reliance on exports of crude oil, most of the region's resource-rich countries have adopted strongly interventionist industrial policies. Particular encouragement has been given to the expansion of petrochemicals, aluminium smelting and iron and steel production. These policies are unlikely to change over the forecast period and will provide a driving force for the region's economic growth.

The Secretariat's energy outlook for the Middle East is based on high annual economic growth, underpinned by high rates of population increase. Population growth is expected to average 3.1 per cent per year, although this is less than the 3.9 per cent achieved between 1980 and 1988. Rising per capita incomes, urbanisation trends and increased demand for private motor transport will all contribute to changes in the regional energy balance.

As noted earlier, the Secretariat's forecasts were prepared prior to developments in the world oil market resulting from the 1990-91 Gulf crisis. In terms of its magnitude and duration, however, the crisis has proved to have had far less impact on world markets than the oil price shock of 1979. At the aggregate level, Middle East production capacity proved sufficient to compensate for most of the drop in Iraqi and Kuwaiti crude output. It can reasonably be expected that these two sources of supply will be re-established in the forecast period. The major impacts of the crisis on the forecasts are likely to be on the direction of trade flows in the short term and on product markets. The loss of 750 thousand barrels per day of Kuwaiti refinery capacity, for example, has highlighted the essential tightness in world refining. This situation has caused some high-demand areas, notably the Asia-Pacific region, to seek alternate product supply sources, as well as to plan increases in domestic refinery capacities.

Total Primary Energy

Energy demand in the Middle East has grown more strongly than any other non-OECD region, averaging 8.5 per cent per year since the early 1970s. The Secretariat's outlook assumes that growth of this magnitude will continue over the period to 2005. In absolute terms, total primary energy supply is expected to almost treble to 666 Mtoe by 2005, compared with 230 Mtoe in 1989.

The Middle East energy balance is heavily weighted towards oil and natural gas, which together represented 98 per cent of total primary energy supply in 1989. The dominance of these two fuels is not expected to decline over the projection period, although there is likely to be an increasing shift towards natural gas. This shift will be induced by deliberate policies to increase the domestic consumption of gas in order to free oil for export markets. Nuclear power has, to date, played no role in the region's energy consumption and this situation is not expected to change. The region's consumption of coal is limited almost exclusively to Israel where its major use is in power generation. In the regional context, coal is unlikely to play an important role. Hydropower potential has been exploited in several countries in the region but it is also unlikely to raise its low contribution to TPES.

As well as absolute energy consumption, energy intensity in the Middle East has risen rapidly since the early 1970s. The currently high level is likely to be maintained as governments continue to pursue growth strategies based on oil- and gas-intensive industrialisation.

Figure 33: **MIDDLE EAST — TPES, 1989-2005**

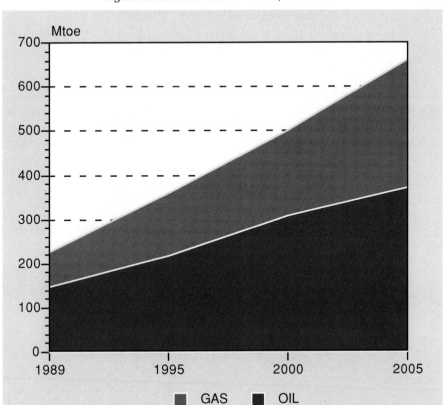

Source: IEA Secretariat

Primary energy production in the region, which is almost exclusively of oil and gas, is expected to grow strongly during the projection period, at an average annual rate of 4.6 per cent. In absolute terms this implies an increase from 923 Mtoe in 1989 to 1897 Mtoe in 2005. As a result, the Middle East is expected to account for 18 per cent of global primary energy production in 2005 compared with 11 per cent in 1989. The region's exportable energy surplus is also expected to increase over the projection period, from 693 Mtoe to 1231 Mtoe, most of which will be accounted for by oil.

Within the region, the five major oil producers — Saudi Arabia, Iran, Iraq, Kuwait and the United Arab Emirates — together accounted for 81 per cent of regional energy demand and 88 per cent of production in 1989. These countries are expected to continue to dominate both energy production and consumption over the outlook period, even given the substantial reconstruction efforts that will be required in Kuwait and Iraq.

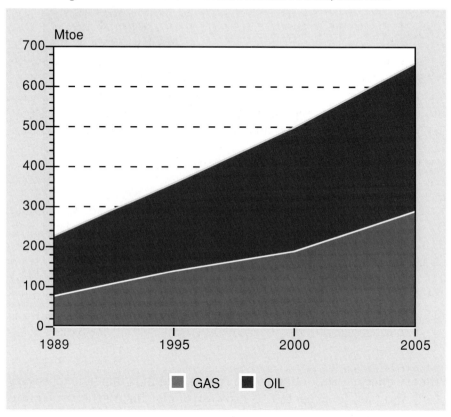

Figure 34: **MIDDLE EAST — ENERGY PRODUCTION, 1989-2005**

Source: IEA Secretariat

Coal

As noted above, the contribution of coal to the region's total energy balance is insignificant. Only in Israel has any significant amount of coal been consumed for electricity generation purposes and its share of that country's total energy demand mix was about 22 per cent in 1989. Overall, regional coal demand is expected to grow at an average annual rate of 1.8 per cent, rising to 4.0 Mtoe in 2005. The dominant share of the increase is expected to come from Israel and will be met almost entirely by imports. The contribution of coal to total regional energy demand will remain insignificant.

Oil

Domestic oil demand in the Middle East has grown more quickly than in any of the non-OECD regions, averaging 8.1 per cent per year since the early 1970s. This trend is expected to continue over the projection period with annual growth of 5.9 per cent. In absolute terms, oil demand is expected to reach 372 Mtoe by 2005, compared with 149 Mtoe in 1989. As well as increased demand from the industry sector, this high growth rate will be underpinned by strong population growth and rising demand for private transportation. Passenger car ownership, for example, is expected to increase by almost 500 per cent, leading to a quadrupling in regional gasoline consumption.

Domestic demand is of relatively minor importance in determining Middle East oil production levels, with export markets playing a far more significant role. There are, however, many considerations that could influence production levels in the region, with the Gulf crisis of 1990 being a forceful reminder of how diverse, unpredictable and far-reaching these can be. As noted earlier, the Secretariat's outlook is based on one of a number of possible oil price assumptions, in particular that crude oil prices begin to rise in real terms as excess production capacity outside the Gulf region is reduced. As this occurs, supply and price will depend on the extent to which Middle East production and ultimately capacity are increased. Underlying the outlook is the existing surplus production capacity in the Middle East and the ability of some producers to expand relatively low-cost capacity quickly. It is also worth noting that 70 per cent of the region's reserves are located in sparsely populated countries where the bulk of easily accessible surplus capacity is located and where the highest reserves to production ratios are found. All these are conditions favouring a long-term view of the oil market and a willingness to protect long-term market prospects by orderly production expansion and moderate pricing. Under this scenario, it is assumed that oil production in the region will grow at an average annual rate of about 4.1 per cent, from 841 Mtoe in 1989 to 1596 Mtoe in 2005.

Proven crude oil reserves in the region were estimated at about 663 billion barrels at the beginning of 1991, equivalent to two-thirds of the global total. About 96 per cent of these reserves lie in the five major Gulf producers — Saudi Arabia, Iraq, the UAE, Kuwait and Iran. At current production levels, these figures give a reserves/production ratio of over 100 years. Combined with the fact that oil production outside the region is expected to remain relatively constant, it is inevitable that the world's dependence on Middle East oil supplies will rise. Under the Secretariat's scenario, the region is expected to account for 38 per cent of world oil production by 2005, compared with 26 per cent in 1989. Its exportable oil surplus is expected to almost double over the same period, from 692 Mtoe to 1223 Mtoe, and an increasing proportion of world incremental oil demand will be met from Middle East supplies.

Natural Gas

Natural gas demand in the Middle East has grown at an average annual rate of about 9.9 per cent since the early 1970s and is expected to continue at about 8.6 per cent over the projection period. In absolute terms, demand is expected to almost quadruple by 2005, reaching 289 Mtoe, compared with 77 Mtoe in 1989. This rapid growth will be underpinned by policies designed to substitute gas for oil in both industry and power generation. Consumption by the industry sector, for example, is expected to increase more than four-fold, and the share of gas in electricity generation is forecast to rise from 40 per cent in 1989 to 75 per cent in 2005. The expansion of natural gas-consuming industrial capacity in the Middle East (as in other gas-producing countries) is viewed as a potential route for indirectly exporting an otherwise almost non-exportable natural resource.

The region's proven reserves of natural gas at the beginning of 1991 were estimated at 1324 trillion cubic feet or 31 per cent of the world total. Iran, the UAE, Saudi Arabia, and Qatar together account for 86 per cent of the regional total. On a global basis, Iran's reserves are second only to those of the Soviet Union. The region's reserves/production ratio is estimated at over 100 years and is even more favourable than its petroleum reserves. Against this background, production of natural gas is assumed to grow rapidly, at an average annual rate of 8.6 per cent to 2005. In absolute terms, production is expected to almost quadruple, to 300 Mtoe, compared with 80 Mtoe in 1988. A large proportion of Middle East gas reserves are found in association with oil and hence gas production will be tied closely to oil output levels.

The region's exportable surplus of natural gas is currently very small when compared with its oil surplus, as most incremental gas production is consumed domestically. In 1989 the surplus was 2.4 Mtoe and is expected to reach 11 Mtoe by 2005. Only two countries currently export outside the region. The UAE exports LNG to Japan, and Iran recently resumed exports to the Soviet Union by pipeline. Iran also intends to export to eastern European countries through the USSR. Although it is currently the largest producer of natural gas in the region, Saudi Arabia is unlikely to generate a significant exportable surplus in the near future as the domestic market appears capable of absorbing increases in production.

Hydropower

As a result of physical conditions, hydroelectric power potential and current supply in the Middle East are insignificant. Hydropower is expected to contribute only 1 Mtoe to total energy supply in 2005, the same as in 1989. Hydro capacity exists in only four countries in the region — Iran, Iraq, Syria and Lebanon. Under an agreement reached between Iran and the Soviet Union in April 1989, the latter is to assist Iran in the construction of a number of hydroelectric facilities, some of which might be realised during the projection period. There are no other definite plans to increase hydro capacity in the region.

Electricity

Urbanisation and rising standards of living, underpinned by increased per capita incomes, will be the driving force for increased electricity demand over the projection period. Electricity demand growth in the region is expected to average 8.2 per cent per year over the period to 2005, implying an absolute increase from 217 TWh to 767 TWh. Fuel inputs to electricity generation are expected to shift increasingly towards natural gas, which will account for about 75 per cent of inputs in 2005, compared with 40 per cent in 1989. Oil's share is expected to decline from 46 per cent to 20 per cent over the same period, reflecting policies to shift demand for oil away from domestic markets towards exports.

Table 9: **Middle East — Primary Energy Balance**[a]
Mtoe

	1989	1995	2000	2005
Coal				
Production	0	1	1	1
Net Imports	3	2	3	3
Consumption	3	3	3	4
Oil				
Production	841	1232	1395	1596
Net Imports	-692	-1013	-1086	-1223
Consumption	149	218	309	372
Gas				
Production	80	146	195	300
Net Imports	-3	-5	-4	-11
Consumption	77	141	191	289
Nuclear	0	0	0	0
Hydro	1	1	1	1
Total				
Production	923	1380	1591	1897
Net Imports	-693	-1016	-1087	-1231
Consumption	230	364	504	666

a   Excludes non-commercial fuels

Note: Because of rounding, totals and sub-totals may not exactly equal the sums of their individual components

Source: IEA Secretariat

USSR

Events since early 1990 in the political and economic life of the USSR make any attempt to forecast the energy outlook fraught with uncertainty. The Secretariat's analysis outlined here was completed prior to the events of August 1991. Nevertheless, it incorporates a cautious approach. The outlook assumes, critically, that the transition to a market economy will occur without any catastrophic disintegration of the Soviet system but that the early part of the outlook period will be marked by economic stagnation.

Enormous changes continue to take place, however, which have increased the uncertainty surrounding the outlook. These have been, principally, an acceleration in the rate of economic decline, a virtual halt in the economic reform programme at a stage where few real gains had been made, and a significant shift in relations between the central government and the republics. The impact of these factors on the energy outlook will be especially strong in the short to medium term. The success or failure of reform efforts will greatly affect energy demand and supply and, hence, the USSR's role in international energy markets, throughout the Secretariat's outlook period to 2005. For these reasons, the following sections place little emphasis on individual numbers and attempt instead to outline in some detail the major risks associated with the forecasts.

The level of economic output in the USSR fell 3 per cent in 1990 and is expected to fall by up to 15 per cent in 1991. At the same time, the inflation rate has climbed sharply, overall investment levels have fallen and external trade has contracted severely. Given the uncertain nature of economic reforms, as well as the current political instability, the deterioration in the main macro-economic aggregates is expected to continue at least into 1992. The major cause for this sharp economic decline is that reforms introduced to dismantle the previous system of central planning have not yet been replaced by effective market-based alternatives. Price increases, for example, which were introduced in the first half of 1991, are considered to have done little to correct fundamental imbalances in relative prices and price controls remain pervasive. As such, little improvement in the allocative efficiency of the Soviet economy has been achieved. Other systemic reforms have also made little progress, including the commercialisation of state enterprises, the establishment of a commercial banking system, development of the private sector and the introduction of foreign exchange policies.

Of particular relevance to the energy sector is the fact that energy prices remain low and have even fallen, limiting incentives for conservation efforts or for increased export performance. Also important for the sector is the current uncertainty surrounding the economic policies of the new sovereign republics and their relationship with the central government. The geographic separation of most energy reserves from domestic customers and export terminals, as well as from equipment manufacturers for the energy industry, leaves the system vulnerable to disruptions in supply and distribution networks if normal economic relations between republics are not maintained. Turmoil in these relations may also hinder the introduction of other reforms, including price liberalisation, on a uniform basis and could impede the inflow of foreign funds for investment purposes. The progress in late 1991 towards the creation of an economic union between a majority of the former republics indicates that this source of disruption may be minimised.

Total Primary Energy

Energy consumption in the Soviet Union grew 3.1 per cent per year over the period 1973-1989 but has since declined as a result of the downturn in economic activity. On the basis that the economic reform programme is successfully implemented, the Secretariat expects energy consumption to rise slowly over the period to 2005. Bearing in mind the declines that have already occurred in consumption, however, most growth is expected to occur in the later years.

The fuel structure of TPES over this period is expected to shift substantially, away from oil towards natural gas, reinforcing the changes which had already occurred over the preceding decade.

Whereas in 1980, 68 per cent of TPES was accounted for by coal and oil, the figure had fallen to 54 per cent in 1989 and is projected to be 42 per cent by 2005. The share of gas is expected to rise from 41 to 50 per cent over the same period with the major increase in total energy consumption expected to come from this fuel source.

Primary energy production is also expected to grow over the period 1989-2005, with most of the growth again restricted to the later years. The USSR is currently the world's largest producer of energy, accounting for 21 per cent of world total primary energy production in 1989, and this share is expected to remain relatively constant. Shifts in the structure of production will be apparent, however, with the share of coal and oil declining from 56 per cent to 40 per cent and

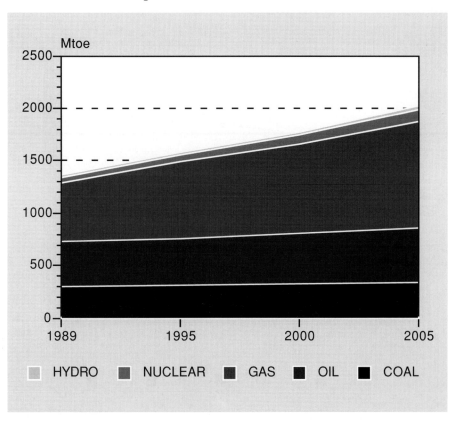

Figure 35: **USSR — TPES, 1989-2005**

Source: IEA Secretariat

that of gas rising from 39 per cent to 54 per cent. Natural gas is expected to contribute almost 90 per cent of the total increase in energy production. These forecasts should be seen against the background of recent falls in energy production. Since 1989, both oil and coal output has declined, and small increases in gas production were insufficient to offset a reduction in total energy output. The risks associated with individual fuel production scenarios are discussed in later sections.

Closely related to future energy production levels is the outlook for energy exports. The USSR's net exports in 1989 were 278 Mtoe and are projected to be around 300 Mtoe in 2005, the bulk of which is

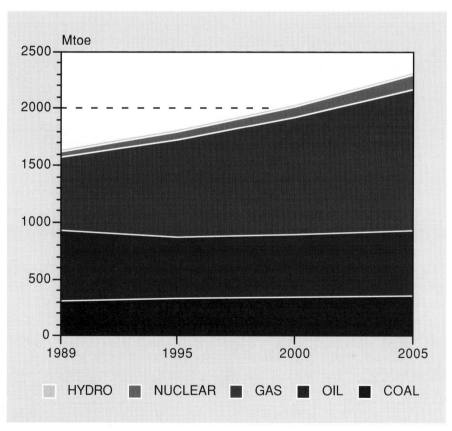

Figure 36: **USSR — ENERGY PRODUCTION, 1989-2005**

Source: IEA Secretariat

likely to come from natural gas. Although the development of the USSR's vast energy reserves will entail enormous difficulties, it is probably not unrealistic to consider that in the longer term, and providing that the economic reform process progresses reasonably, these reserves may provide an increased share of the western world's energy requirements.

One of the major features of the Soviet energy economy of recent years has been its persistently high energy intensity. In 1989, energy intensity (in constant 1985 $ US) was 0.99, compared with an average of 0.41 for IEA countries. Soviet intensity has deteriorated since the early 1960s. Reasons for this situation include the dominance of heavy industry in the industrial structure and the systemic difficulties of achieving energy savings. Most importantly, the absence of a market price mechanism has provided little incentive for energy conservation. High energy intensity is likely to be a continuing characteristic of the economy until economic, and especially price, reforms are effectively implemented. Even if these reforms are successful, the decline in energy intensity will probably not be as rapid as that achieved in OECD countries during the 1970s and the early 1980s. The lower and falling intensities in IEA countries over that period reflect, in part, the capital-intensive improvements in energy-efficient equipment and technology made in response to the sharp rise in world energy prices, as well as shifts in the industrial structure towards services and high-technology sectors. Economic structures requiring less energy consumption will only emerge slowly in the USSR. In addition, successful transition to a market economy will raise private purchasing power and will lead to increasing demand for energy-consuming commodities such as private cars, home heating and electrical appliances. Overall, the Secretariat assumes that energy intensity in the USSR will rise through the mid-1990s before declining thereafter.

Coal

Despite substantial reserves, coal is expected to play a declining role in the Soviet fuel balance. Its share of TPES is projected to fall from 22 per cent in 1989 to 17 per cent by 2005 and the absolute level of coal consumption is expected to be only slightly higher in 2005 than it was in 1989. Coal production is projected to follow a similar pattern, with average annual increases of less than 1 per cent over the period to 2005. Net exports of coal are expected to remain small.

According to Soviet estimates, total proven recoverable coal reserves amount to some 240 billion metric tons, equal to about 45 per cent of the world total. These reserves include 140 billion metric tons of hard coal (anthracite, bituminous and sub-bituminous) and about 100 billion metric tons of lignite. The production potential is, therefore, substantial, but there are important constraints on coal development. The quality and accessibility of reserves, for example, is generally declining, particularly in the western (European) part of the country. About 75 per cent of recoverable reserves are now located in the Asian USSR and tend to be far from the major industrial and population centres. This entails difficult and costly transport and distribution requirements, in a system where infrastructure is inadequate and rolling stock in short supply. This situation has lead to the development of some coal-fired mine-mouth power plants in the eastern regions but distribution losses in these circumstances are high. In addition, coal in these regions is generally of lower quality, suffering from high water, ash and sulphur levels. The coal industry is also characterised by obsolete and badly maintained equipment and machinery and is in need of investment for replacement, repair and expansion. As coal exports have not been a large earner of foreign exchange, however, the industry has a low priority for scarce investment funds and, hence, efficiency gains will be difficult to achieve. In the medium term, it can be expected that coal inputs to electricity generation will be displaced increasingly by gas.

Oil

The Secretariat estimates that Soviet oil demand will rise over the period between 1989 and 2005, although more slowly than between 1973 and 1989. The lower growth in consumption will be, in part, a response to the anticipated lower rates of economic growth over the early forecast period, but also a result of policy initiatives to shift domestic demand away from oil towards gas. Pricing reforms will be important in determining the pace of such a shift.

The extent to which the substitution of gas for oil can occur over the forecast period will also be determined in part by the structure of the refining sector. About 40 per cent of current refinery output is heavy products, mainly for power generation. The current shift in demand away from fuel oil towards lighter fuels, especially for transport, will require upgrading of the present refinery configuration. The only way that refineries can presently meet the demand for light fuels is by running more crude, increasing heavy fuel output in the process. If, at the same time, demand for heavy fuel oil in the power sector falls as a result of increased gas consumption, the excess fuel production will be magnified. Substantial investments, particularly in catalytic cracking capacity, would be required to avoid this situation. Conversely, improvements in the refinery system would result in lower domestic demand for crude oil, freeing any excess for the export market.

Soviet oil production in the Secretariat's outlook is expected to fall between 1989 and 1995 but to recover to some extent from the second half of the 1990s. The expected decline in production in the early years, which continues a trend evident since 1988, is attributable to the gradual depletion of known resources combined with the increasing cost of developing new reserves. Outside Western Siberia, most oil fields have reached maturity and are in steady decline. Even some of the major Western Siberian fields, from which two-thirds of Soviet oil is sourced, are considered to have passed their physical peak and total proven reserves are now considered equal to about 12 to 14 years' production. While new discoveries are reportedly being made at a rate greater than production they are generally in small, remote fields in the east which are significantly more difficult and costly to develop. To sustain even current levels of oil output will require large increases in investment in the sector, both to replace the frequently low-quality and outdated equipment and to increase exploratory drilling rates. Given the current shifts in investment priorities for central funds mentioned above, it is unlikely that additional autofinancing and bank credits can make up for the shortfall. However, the oil sector has been historically a key to industrial growth as well as an important foreign exchange earner and it would be reasonable to assume that some priority will be given to sustaining levels of investment. The alternative is to put greater reliance on foreign investment, particularly joint ventures with Western partners as a source of equipment and expertise. While this poses significant administrative and technical problems in the short term, it is possible that by the second half of the decade such arrangements may result in some stabilisation, if not increases, in Soviet oil output. An important consideration in the longer term (possibly beyond the Secretariat's outlook period) is the fact that the USSR remains a relatively immature oil region, with vast unexplored territory. It is possible that significant reserves remain undiscovered and it is for this reason that foreign interest has been maintained in hydrocarbon exploration and development.

The USSR is expected to remain a small net exporter of oil but the exportable surplus is projected to decline. Since substantial increases in production are unlikely, the ability or willingness of the Soviet Union to continue to supply oil to the export market will depend to a large extent on the success of policies to reduce domestic oil demand. Following the introduction in 1991 of hard-currency trade with eastern Europe, Soviet oil deliveries to this area are expected to fall.

Natural Gas

The component of the Soviet energy balance expected to show the strongest growth is natural gas. Over the period to 2005, gas is expected to account for about 70 per cent of the total increase in Soviet energy consumption. As a consequence, the share of gas in TPES is expected to rise from 41 per cent to 50 per cent over the same period. The major gains will continue to be made in industry and power generation, where concerted efforts have been made to substitute gas for fuel oil in order to free oil for export. Increased use of gas in the power sector is also planned as a partial replacement for the expected slowdown in the nuclear programme. Some may also be used to replace coal for combustion in polluted areas.

Production growth is projected to continue and, as a proportion of total energy production, natural gas is expected to rise from 39 per cent in 1989 to 54 per cent in 2005. The exportable gas surplus is forecast to increase from 89 Mtoe in 1989 to about 227 Mtoe in 2005.

The Soviet Union's proven recoverable gas reserves at the beginning of 1991 are estimated at 1600 trillion cubic feet, equal to 38 per cent of total world reserves, and providing a reserves/production ratio of

56 years. The majority of the reserves are in the northern regions of Western Siberia, an area in which both exploration and exploitation of reserves are relatively difficult. The Ukraine, Volga-Urals and Central Asian basins are subject to declining production levels, in part because greater financial and technical resources are being devoted to the giant Western Siberian fields. Recent discoveries of offshore gas in the Barents and Kara Seas, however, have provided substantial production prospects in the medium to longer term, although significant technical and financial difficulties may necessitate the inclusion of foreign partners in the exploitation of these reserves. Offshore production is expected to commence in the Okhotsk Sea in the near future.

With such vast reserves it is inevitable that natural gas will play a major role in the growth of Soviet energy consumption, particularly as opposition to nuclear power increases and the foreign exchange earning capacity of the oil industry is increasingly exploited. The sector suffers, however, from many of the same problems as the oil industry, namely the increasingly remote location of new reserves, low-quality and outdated equipment and an inadequate distribution and storage network. It is only the low well-head costs which have, to date, disguised the effects of these factors. If gas is to increase its dominance at the projected rate, certain constraints will need to be overcome, particularly in the transport and distribution areas where infrastructure has often deteriorated due to hasty construction and inadequate maintenance. Transport and distribution issues will become increasingly difficult as recent gas discoveries have tended to be in more distant locations. The production of natural gas is currently demand rather than supply constrained and the absorption of increased gas supplies will require the construction of additional storage and electricity generation facilities, as well as increased gas-using equipment such as central heating in cities. Given the volume of reserves, the potential to increase gas exports without causing domestic shortages is vast.

Nuclear

Soviet nuclear output is projected to more than double over the period to 2005, although the base is small relative to other fuels.

Nuclear power is expected to account for 10 per cent of the total increase in energy consumption over the period and its share of TPES is projected to rise from 4 to 6 per cent.

The likelihood of these projections being met will depend in large part on the extent to which the growing public sensitivity to nuclear power is taken into consideration in nuclear policy-making. In the post-Chernobyl era the nuclear programme has been in a state of flux with a series of cancellations or postponements of new projects, leading to a serious falling behind of targets. Indeed, only 10 out of a proposed 41 GW which were envisaged in the 1985-1990 five year plan have been brought on line. Nuclear operating capacity at end-1990 stood at 35 GW and a further 19 GW were under construction. With nuclear power plants providing about 12.5 per cent of electricity supply and a large part of generating capacity under construction, significant plant closures could exacerbate electricity shortages and lead to further economic disruption. Cancellation of reactors under construction could require the building of alternative replacement capacity with implications for fossil fuel demand, particularly of natural gas. It would also limit the options available for reducing fossil fuel-related emissions of atmospheric pollutants and greenhouse gases.

Hydropower

Hydropower generation is expected to increase between 1989 and 2005 although its share of TPES will remain at about 2 per cent throughout the period and its share of electricity output at about 14 per cent.

While there is substantial hydropower potential in the USSR (up to about 270 GW), about two-thirds is located in the eastern part of the country, away from the major load centres. Soviet planners estimate that installed hydroelectric capacity could rise to 90 GW by the turn of the century, from 67 GW at end-1989. Further development of hydropower is likely to face significant opposition, however, based on economics, environmental considerations and socio-economic impacts, including the flooding of agricultural land and the displacement of villages.

Electricity

Growth in total electricity consumption is expected to average 2.6 per cent per year over the period 1989-2005 with the majority of the increase occurring in the later years. In 2005, almost 70 per cent of electricity is expected to be generated from conventional thermal sources. This is little different from the situation in 1989, although greater reliance will be placed on natural gas. Since 1980, there has been a clear move away from oil towards coal and natural gas for power generation purposes — no new oil-fired plants have been constructed and some existing plants have been converted to gas. While coal-fired plants will continue to expand, most of the growth in output will be accounted for by increases in gas-fired plant.

An important factor complicating the USSR's electricity system is the enormous distances between fuel reserves (primarily coal and gas) in the east and the major demand centres in the west. Fuel must be shipped to generating facilities in the west or the electricity generated in the east must be transmitted, with concomitant losses, to the west. The choice between the siting of electricity generation facilities near raw material supplies versus near the major load centres, and the different transportation demands these entail, will continue to be an important factor in investment planning for the electricity sector. Currently, the electricity grid is relatively old and not well maintained. It is also stretched in terms of carrying capacity which leads to an increasing number of unplanned outages. In the absence of substantial investment, shortages in generating capacity may occur in the future, mainly in winter periods of high demand and in certain regions. The primary impact of such shortages will be borne by the industrial sector as residential consumers are generally accorded higher priority.

Table 10: **USSR — Primary Energy Balance**[a]
Mtoe

	1989	2005
Coal		
Production	314	357
Net Imports	-12	-23
Consumption	301	334
Oil		
Production	610	571
Net Imports	-177	-50
Consumption	433	521
Gas		
Production	644	1246
Net Imports	-89	-227
Consumption	555	1019
Nuclear	55	120
Hydro	21	29
Total		
Production	1644	2323
Net Imports	-278	-300
Consumption	1366	2023

a   Excludes non-commercial fuels

Note: Because of rounding, totals and sub-totals may not exactly equal the sums of their individual components

Source: IEA Secretariat

V. ENERGY-RELATED ENVIRONMENTAL ISSUES IN NON-OECD COUNTRIES

THE INTERNATIONAL DIMENSION

The local and regional impacts of energy activities on the environment have been the focus of policy attention in OECD member countries for at least the last two decades. More recently, the existence of similar problems has begun to be recognised in some non-member countries, usually in response to the environmental pressures exerted by population growth and the processes of industrialisation and urbanisation. The problems vary enormously from one region to another, depending largely on the level and structure of economic development, and range from deforestation associated with fuel-wood consumption in sub-Saharan Africa to air and water quality and acid rain problems in the industrial economies of Asia and eastern Europe. As a result, the environmental problem has become widespread — issues which were once the domain of the developed industrialised world have now assumed much broader geographic significance.

In addition to this trend is the fact that some of the more pressing environmental problems are no longer local, regional or even national in their impacts but have become transboundary in nature. There are essentially two aspects to the transboundary issue. The first is the conceptually simple case where the impacts of pollution are borne by countries other than the polluter, with acid rain being a significant example. The transboundary effects in these cases are generally limited to neighbouring or nearby countries and solutions can be implemented at a local or regional level. The second case encompasses the more far-reaching, global impacts of any one country's actions. The prime examples are the emission of greenhouse gases and their potential impact on global climate change and the problem of stratospheric ozone depletion resulting from the emissions of chlorofluorocarbons (CFCs) and related substances. Fundamental to an analysis of this case is the fact that the global nature of the environmental system means that the consequences of any one country's actions will be felt by all other countries. The intensity of the impacts of these actions, however, may vary across countries according to individual circumstances. Low-lying or island countries, for example, especially those with limited financial resources, will probably be far more affected by global climate change than landlocked developed countries. Effective solutions to global problems will require extensive multilateral co-operation and recognition of the inter-related interests of countries at very different stages of development and with vastly divergent access to resources.

The important role of energy activities in environmental degradation is well documented. The combustion of fossil fuels, for example, is a major source of sulphur dioxide (SO_2), nitrogen oxides (NO_X) and other conventional pollutants, as well as making a substantial contribution to global increases of carbon dioxide emissions, the major contributor to the greenhouse effect. The increasingly significant share of non-OECD member countries in global energy consumption is, therefore, likely to result in their being responsible for an increasing share of global environmental problems. The non-OECD world is expected to account for more than half of global energy consumption by the mid-1990s and for about 57 per cent of consumption in 2005, compared with 49 per cent in 1989. Its significance will also increase because, for a variety of financial and technical reasons, action to ameliorate environmental problems is likely to be pursued more vigorously in OECD countries than in non-members. The transboundary and global nature of environmental problems, however, will ensure that these problems remain well within the domain of OECD action and concern. It also points to an area where increasing international co-operation will be required if effective global solutions are to be devised.

AREAS OF ENVIRONMENTAL CONCERN

The IEA has previously identified major areas of environmental concern where energy activities can play an important role[1]. These are:

- major environmental accidents;
- water and maritime pollution;
- land use and siting impact;
- radiation and radioactivity;
- solid waste disposal;
- hazardous air pollutants;
- ambient air quality;
- acid deposition;
- stratospheric ozone depletion; and
- global climate change.

Most of these areas of concern will, to some extent, apply to all the world's regions, although their impact and relative importance will vary according to the nature of the energy activities in any region and the degree of environmental controls. The current priority afforded to resolving environmental problems also varies significantly across countries and is largely a function of the severity, or the perceived severity, of the environmental impacts of energy activities, as well as the availability of financial resources. In most non-OECD regions some efforts have been made to curtail the environmental impacts of energy-related activities and to repair the consequences of past damage. In general, however, it would be reasonable to conclude that the concerns of non-member countries continue to be concentrated at local level problems, such as urban air pollution. Transboundary and global problems have, to date, received little direct policy attention.

In China, energy-related environmental problems largely reflect the dominance of coal in the national energy balance. Coal is the major fuel source in both the household and industry sectors, as well as for

---

1. International Energy Agency [IEA] (1989). *Energy and the Environment: Policy Overview.* OECD, Paris.

electricity generation. Coal consumption in all sectors is characterised by inefficient combustion technologies and inadequate emissions control equipment. In the household sector, for example, average efficiencies as low as 12 per cent have been cited for cooking and heating appliances. The environmental problems associated with electricity generation have been compounded by the recent policy of constructing very small coal-fired plants to supplement local electricity supplies. These are usually less efficient and more polluting than larger plants built at the provincial and state level. The consequences of these factors are high levels of urban pollution and an increasing incidence of acid precipitation, both of which are exacerbated by the high-ash and high-sulphur content of most domestically consumed coal. The second important group of energy-related environmental problems in China are those associated with the generation of hydroelectricity. Hydropower is the only significant alternative to coal-fired electricity and accounts for about 22 per cent of electricity generation. A substantial proportion of this is sourced from small, local power stations, although large projects are also important. The environmental problems associated with hydro developments include salinisation of river deltas, erosion and siltation of dams and the alienation of productive farmland. In the USSR, the energy sector is a major contributor to many of the worst environmental and safety problems in that country. Low ambient air quality, a spate of serious and environmentally costly accidents at energy facilities, extensive marine and water pollution, the degradation of useful lands, and acid rain are notable examples. The extensive production and use of coal, in particular, is responsible for substantial land alienation and emissions of SO_2, NO_X and other conventional pollutants. Attempts to ameliorate urban air quality by the use of tall exhaust stacks from power plants have actually worsened the acid rain problem by dispersing pollutants across a larger geographic region. Energy-related environmental concerns in the economies of eastern Europe are also largely coal-based and include conventional urban air pollution as well as large tracts of land affected by acid rain. Urban air quality is also probably the primary concern in the densely populated centres of the Asia-Pacific region and in the Middle East and Latin America. Coal- and oil-based electricity generation sectors, heavy industry and rapidly growing transport requirements have all contributed to conventional pollution problems in these regions. In addition, concerns with the environmental consequences of hydropower developments are being felt to an increasing degree in Latin America. In the African region and parts of south Asia, the predominant environmental problem in the energy sector is deforestation associated with the consumption of fuel woods. The pressures of population growth combined with inefficient land management practices have resulted in widespread land degradation, compounding the fuel shortage problem.

AN OVERVIEW OF THE MAJOR TRANSBOUNDARY AND GLOBAL ENVIRONMENTAL PROBLEMS

Given that the concerns of this report are, primarily, to analyse the implications for OECD member countries of developments in the energy situation of the non-member world, only those areas of environmental concern with potential transboundary or global implications are treated in any detail here. The most important of these are acid deposition, stratospheric ozone depletion and global climate change, although the siting of energy activities close to international boundaries, major environmental accidents, particularly those which are nuclear-related, and maritime pollution may also have an international dimension.

Acid Deposition

The major cause of acid deposition has been found to be emissions of SO_2 and NO_X. These pollutants not only cause local air pollution but also contribute to the regional and transboundary problem of acid rain. Volatile organic compounds (VOC), chlorides and ozone also probably participate in the complex set of chemical transformations in the atmosphere that result in acid deposition and the formation of other regional air pollutants. The precise relationship between emissions and levels of damage is still uncertain, although acid deposition is considered to have a wide range of environmental effects. These can be spread over large areas due to the long-distance transport of the pollutants involved. The most commonly cited effects include:

- acidification of lakes, streams and groundwaters, resulting in damage to fish and other aquatic life;

- damage to forests and sometimes to agricultural crops; and
- deterioration of man-made materials, such as buildings and metal structures.

Energy-related activities are major sources of the main identified precursors of acid deposition. Electric power stations, residential heating and industrial energy use are estimated to account for 80 per cent of total anthropogenic SO_2 emissions (with coal alone accounting for 70 per cent), and road transport is an important source of NO_X emissions. One of the simplest and cheapest methods of dealing with the environmental impacts of emissions at the local level — the dispersal of industrial and power plant emissions from tall stacks — facilitates the transport of these gases over long distances. Once in the atmosphere, SO_2 and NO_X are readily converted into sulphuric and nitric acids, bringing acid deposition to non-source regions which, in some cases, may be beyond national boundaries.

While a systematic global assessment of urban air quality (and, hence, the potential for acid deposition problems) is not available, the United Nations Environment Programme[1] (UNEP) has measured SO_2 levels over the period 1973 to 1984 in a sample of 33 cities, including eleven outside the OECD region. Of this group of eleven, seven were shown to have increasing concentrations of SO_2, while levels in the remaining four had either remained constant or declined. This compares with only one out of 22 OECD cities where SO_2 concentrations had risen. The same disproportionately deteriorating trends are evident in non-OECD cities compared with the OECD when measures of the peak concentrations of SO_2 are observed. Out of 54 cities measured between 1980 and 1984, 27 were found to have absolute levels of SO_2 which exceeded World Health Organisation (WHO) guidelines on more than seven days per year, 16 of which were in the non-OECD. Other data (see Table 11) indicate the absolute emission levels of SO_X (principally SO_2) and NO_X in selected Asian and OECD countries.

---

1. United Nations Environment Programme [UNEP] and World Health Organization [WHO] (1988). *Global Environmental Monitoring System — Assessment of Urban Air Quality.* UNEP/WHO, London.

Given the high proportion of global man-made SO_2 emissions originating from the energy sector, there is a strong likelihood that concentrations in non-OECD countries will increase as consumption of fossil fuels continues to rise. This trend will be reinforced, at least for a number of decades, because of the recent and continuing tightening of emission control regulations applying to both power plants and diesel fuels in, for example, the United States and Europe. Other factors are also likely to contribute to increasing disparities between the two groups of countries, including the continuing shift to low-sulphur coals in the OECD and the already greater reliance on high-sulphur fuels in non-OECD countries, especially eastern Europe and China. The lower accessibility of developing countries to advanced technologies for pre- or post-combustion sulphur removal will also limit their capacity to respond.

Emissions of nitrogen oxides are more difficult to measure than those of SO_2 and hence there is even less systematic data available, especially on a trend basis. Fossil fuel combustion is estimated to account for about 75 per cent of NO_X emissions, divided equally between stationary sources, such as power stations, and mobile vehicular sources. UNEP data indicate that urban NO_X levels are probably increasing in the rapidly industrialising countries, especially where vehicle numbers are increasing. Cities where evidence of such a trend was found include Sao Paulo, New Delhi, Hong Kong, Bombay and Santiago. In contrast, Singapore's levels were found to have declined markedly over the ten years to 1985, probably because of increasing controls on motor vehicle emissions. The problem is certainly not confined to non-OECD countries. The data indicate an upward trend in several of the larger European cities which were measured, including London, Frankfurt and Amsterdam. This may prove to be a short-term trend, however, as the impact of catalytic converters, the principal control technology, will continue to increase as they are made compulsory in more European countries. Even in those countries where catalytic converters are already required technology they have yet to reach saturation point.

While the measurement of SO_2 and NO_X emissions and atmospheric concentrations is difficult, it is much more so to measure and monitor the migration and deposition of pollutants. Acid precipitation has been recognised as an environmental problem in

Table 11: **SO$_x$ and NO$_x$ Emissions in Selected Asian and OECD Countries**
1987
(1000 metric tons/year)

	SO$_x$	NO$_x$
China	16470	3910
India	3760	1770
South Korea	1010	440
Indonesia	820	310
Thailand	720	290
Taiwan	620	200
Philippines	440	90
Malaysia	320	130
North Korea	280	250
Singapore	210	70
Denmark	248	266
Finland	-	260
France	1517	1652
Norway	93	232
Sweden	220	300
United Kingdom	3867	2303
Japan	2778	1986

Sources:
1) For Asian data (preliminary) including Japan: National Institute of Science and Technology Policy, Science and Technology Agency of Japan
2) For OECD data, excluding Japan: OECD Environmental Data: Compendium 1989

Scandinavia for several decades but the geographic area now perceived as threatened has expanded far beyond these boundaries to encompass almost all of Europe (including the European part of the Soviet Union), as well as parts of North America and some industrialised areas of the developing world, including China and Brazil. In Europe, the largest amount of acid deposition is occurring in Germany, Poland, Czechoslovakia, Italy and Austria, with Poland experiencing the highest rate of all. Of further interest are figures indicating the proportion of acid deposition which originates from outside the country concerned. As Table 12, based on 1980 data, indicates, this proportion was as high as 92 per cent in Norway, followed by 90 per cent in Switzerland and 85 per cent in Austria, although the eastern European countries were also the recipients of transboundary acid pollutants with 63 per cent of acid deposition in Czechoslovakia sourced externally and 58 per cent in Poland.

In China, considerable evidence of acid deposition has been found in various areas of the country and is usually associated with the combustion of coal and other fuels in urban and industrial regions. It is considered likely that acid and particulate pollutants are transported from China to Japan and as far as Hawaii. Although the northern hemisphere is responsible for the larger proportion of global acid rain problems, the phenomenon is not unknown in the southern hemisphere. In general, pollution control requirements are less enforced in some Latin American and many African countries. As a result, there are a number of local and regional deposition problems downwind from major urban and industrial centres but, as the level of economic activity and energy consumption is lower, there are few problems on the geographic scale of the northern hemisphere.

Table 12: **Levels and Sources of Sulphur Deposition in Selected European Countries**

	Average Monthly Deposition (100 metric tons)	Percentage Deposition Received from Other Countries
Czechoslovakia	1301	63
Eastern Germany[1]	778	36
Poland	1330	58
Austria	341	85
Belgium	161	58
Denmark	109	64
France	121	48
Western Germany[2]	1158	52
Italy	1132	30
Netherlands	173	77
Norway	255	92
Sweden	475	82
Switzerland	141	90
UK	847	20

1   GDR before the unification of Germany
2   FRG before the unification of Germany

Source: Baldwin, I. "Acid Rain: A Global Perspective", The Environment Professional, (7: 3) 1985

For a number of reasons the problem of acid deposition and its potential transboundary implications are likely to become of increasing importance in the non-OECD industrialising economies. On a global scale, the problem is also likely to become increasingly concentrated in non-OECD regions, primarily as a result of the higher rates of growth in energy consumption which will be experienced in these areas compared with OECD countries. In addition, emission control requirements, which have been shown to have a positive effect on emission trends, particularly for SO_2, are more likely to be enforced in OECD countries. In the case of NO_X, emission controls have been successfully applied in some countries, notably the United States and Japan, but absolute OECD emissions have continued to rise because of increases in vehicle numbers. This effect is likely to be exacerbated in the non-OECD world where the pressures of population growth and economic development will inevitably lead to increased demand for motor transportation. A further factor complicating the outlook in some non-OECD regions, particularly eastern Europe and China, is the large indigenous reserves of high-sulphur coal which, when combined with shortages of hard currency, will act to constrain fuel substitution possibilities.

The transboundary implications of acid rain and the need for international co-operation to resolve such environmental problems have been recognised in the international negotiating framework, resulting in a convention on Long-Range Transboundary Air Pollution (LRTAP). This was established by the United Nations Economic Commission on Europe (UN-ECE) and was joined in 1983 by 30 countries. The convention requires member countries to exchange information on acid precipitation and to consider its transboundary effects in their development planning but imposes no requirement to reduce emissions. Eighteen member countries (including Bulgaria, Czechoslovakia, Hungary and the USSR) signed a protocol to the convention on SO_2 which calls for emission reductions of 30 per cent from the 1980 level by 1993. The protocol was adopted in 1985 and came into force in 1987. A second protocol was signed in 1988 by 24 countries (including, from outside the OECD, those mentioned above plus Poland and the former eastern Germany) which agreed to limit NO_X emissions to their 1987 level by end-1994. The NO_X protocol will come into force when it has been ratified by a sufficient number of countries.

Stratospheric Ozone Depletion

Stratospheric ozone depletion and its regional distortion is a global environmental problem caused by emissions of chlorofluorocarbons (CFCs), halons and nitrous oxide (N_2O). The effect of ozone destruction is to allow an increase in the amount of ultraviolet radiation reaching the earth's surface which has potentially harmful effects on human health as well as on the productivity of terrestrial and aquatic ecosystems. Depletion of stratospheric ozone occurs when free chlorine atoms, resulting from the breakdown of CFCs and related substances, react with ozone to form chlorine oxide and oxygen. (Natural sources of stratospheric chlorine also exist, for example chloromethane given off by rotting vegetation, but these are now a minor source compared with the man-made compounds). A catalytic cycle of ozone destruction is initiated as the chlorine oxide is, in turn, broken up, allowing the chlorine atom to again react with ozone. In this manner, a single chlorine atom can be responsible for the destruction of thousands of ozone molecules. A similar catalytic reaction can be initiated by oxides of nitrogen. Measurements undertaken by a group of international bodies, including UNEP and the World Meteorological Organisation, have shown a significant long-term reduction in ozone levels when analysed over specific latitude bands and by season. A substantial decline in the ozone layer has, for example, been measured over Antarctica, particularly in the months of September and October.

Energy activities are only partially responsible (directly or indirectly) for emissions of CFCs and N_2O. Fossil fuel and biomass combustion are responsible for a significant proportion of anthropogenic N_2O emissions but it is CFCs which play by far the more important role in ozone depletion. The main energy-related sources of CFCs are those used as refrigerants in transport and space conditioning and refrigeration equipment, or as blowing agents in foam insulation. These applications account for about 60 per cent of CFC uses. Although only a minor proportion of the ozone depletion problem is directly attributable to fuel combustion, it is considered an energy-related issue because the major uses of CFCs are energy-intensive activities.

It is estimated that about 1.1 million metric tons of CFCs were produced globally in 1986 and that the growth in consumption has been substantial. The most important users of CFCs are the industrialised countries, with North America and western Europe accounting for roughly two-thirds of global consumption. While the non-OECD countries still play a minor role in global CFC consumption, particularly when measured on a per capita basis, there is considerable potential for their use in this region to increase. As population and income levels rise, demand for CFC-containing products, particularly in the areas of refrigeration and space conditioning, will certainly increase. It is in this context that international negotiations to limit the production and use of CFCs have been, and will continue to be, important.

Significant progress in this regard has already been made in the international arena under the umbrella of the Vienna Convention for the Protection of the Ozone Layer. The objectives of the convention are to protect human health and the environment against adverse effects resulting from modifications to the ozone layer. Negotiated under the aegis of UNEP, the Vienna Convention, which was adopted in 1985 and came into force in 1988, recognised the future need to develop specific regulatory protocols which would be binding on parties to the agreement. Over 30 countries had signed the convention by early 1989 and others have followed. A protocol to the Vienna Convention was negotiated in Montreal in 1987 and came into force on 1 January 1989. It represents a specific agreement by the world's prime producers and consumers to reduce their atmospheric emissions of the five major CFCs and three related compounds. As originally drafted, the Montreal Protocol committed signatory nations to reduce their production and consumption of CFCs to half their 1986 level by 1998. Production was subject to less severe cutbacks with the intention of meeting the growing demands from developing countries with no access to alternatives. Developing country interests were further addressed by their exemption from the consumption provisions for a period of ten years, provided that their annual per capita consumption of the regulated substances is less than 0.3 kilograms. The protocol also imposes restrictions on the import and export of the restricted substances between countries that are parties to the protocol and those that are not.

By mid-1991 more than 60 countries, together accounting for about 85 per cent of global CFC consumption, had either ratified the protocol or announced their intentions to do so. These included all OECD countries plus the USSR, Thailand, Mexico, Venezuela, Argentina, Panama, Egypt, Nigeria, Ghana and Uganda. The notable exceptions are China, India and Brazil, the major non-OECD producers and consumers of CFCs. In 1989, a group of countries, including the US and some EC states, announced intentions to go beyond the requirements of the protocol. This initiative resulted in a 1990 amendment to the protocol requiring the elimination of CFCs in the developed world by 2000 and in developing countries by 2010 and its extension to include all CFCs and halons.

An important outcome of the 1990 initiative was an agreement by developed nations to establish a global ozone-protection fund. The fund, to be administered jointly by the World Bank, UNEP and the United Nations Development Programme (UNDP), is intended to facilitate the transfer of ozone-friendly technology to developing countries[1]. Two of the major recipients of the assistance are expected to be China and India, as a result of which it is anticipated that both nations will ratify the agreement.

The Montreal Protocol is significant as it represents a test case of the world's ability to deal with intractable global environmental problems. Its successful implementation has demonstrated that the international community is capable of negotiating co-operative agreements for the benefit of all countries as well as of future generations. It is also the first international attempt to bind large developing countries such as China and India to potentially costly environmental accords. In addition, it is the first time that an international environmental agreement has been backed by substantial funding and its principles may be used to establish a framework for future international exchanges of financial aid. The successful implementation of the Montreal Protocol could be seen as a model for dealing with other environmental issues of an international or global nature, including the problem of global climate change. If so, the financial consequences for OECD countries could well be substantial.

Greenhouse Gas Emissions and the Problem of Global Climate Change

The concern about global climate change resulting from the effect of atmospheric concentrations of greenhouse gases is potentially the

---

1. The fund has since been incorporated into the same three agencies' Global Environment Facility (GEF), expected to become operational in 1991. This is a three-year pilot programme which will provide financing of about $1.5 billion to support innovative projects and programmes affecting the global environment. The GEF's four areas of operation are protecting the ozone layer, limiting greenhouse gas emissions, protecting biodiversity and protecting international waters. Resources allocated under the ozone protection area will be kept separate from the GEF's core fund and funding under this facility will apply only to countries that are signatories to the Montreal Protocol.

most important emerging environmental problem relating to energy. The most significant of the greenhouse gases are carbon dioxide (CO_2), methane (CH_4), nitrous oxide (N_2O), tropospheric ozone (O_3) and CFCs. Carbon monoxide (CO), NO_X and non-methane VOC are also important as they directly influence the concentrations of other greenhouse gases in the atmosphere. At present, it is estimated that CO_2 contributes about 50 per cent to the anthropogenic greenhouse effect, with methane and CFCs each contributing about 15 per cent and N_2O about 9 per cent. Energy activities play an important role in the release of anthropogenic greenhouse gases:

- fossil fuel burning accounts for about 75 per cent of global anthropogenic CO_2 released, the remainder coming mainly from deforestation and oxidation of exposed soil;

- combustion of fossil fuels and biomass together account for a substantial proportion of anthropogenic emissions of N_2O;

- ozone is the product of reactions involving pollutants from fossil fuel use (mainly NO_X and VOC);

- in the case of methane, most emissions are due to the anaerobic decomposition of organic matter. Fuel consumption and distribution systems, principally of natural gas, may account for 10 to 30 per cent of total emissions. Methane is also released during the mining of coal.

As well as accounting for a significant proportion of total greenhouse gas emissions, emissions of CO_2 are relatively methodologically easy to estimate as they are directly related to the combustion of fossil fuels. Estimation of other greenhouse gas emissions is more complicated and imposes much larger data requirements. Emissions of CH_4 from leakages in the natural gas, coal and oil fuel cycles, for example, vary not only from country to country, but from one production zone to another. Current estimates of CH_4 emission rates from all sectors, including the energy sector, vary significantly. Unlike CO_2 and CH_4, NO_X, N_2O and CO emissions are largely dependent on combustion and atmospheric conditions such as fuel/air ratios and temperatures, and are influenced substantially by emission control technologies. The data set required to estimate emissions of these gases is much larger than that for CO_2. For this reason and the fact that CO_2 is the primary contributor to the anthropogenic greenhouse effect, the following analysis has been limited to an assessment of the extent of CO_2 emissions from non-OECD regions and a comparison with those from the OECD.

Given that TPES growth rates in the non-OECD world were in excess of 4 per cent per year between 1973 and 1989, it is not surprising that energy-related CO_2 emissions have also grown strongly. Preliminary estimates indicate that emissions of CO_2 from fossil fuel combustion grew at an annual rate of 3.6 per cent, from 1751 million metric tons of carbon (mtc) in 1973 to 3102 mtc in 1989. If traditional fuels are included in the analysis, the figures rise to 2038 mtc in 1973 and 3483 mtc in 1989. Changes in the fuel structure of non-OECD TPES have been important in containing the growth rate of carbon emissions — this has been less than that of energy demand as a result of the increasing share of natural gas and non-fossil fuels in the total energy mix. In contrast with the non-OECD, emissions of CO_2 from the combustion of fossil fuels in OECD countries rose at an annual rate of only 0.4 per cent over the same period and, as a result, the non-OECD's share of world emissions increased from 40 per cent in 1973 to 53 per cent in 1989. In a global context it is equally striking that, over the same time period, the non-OECD countries were responsible for almost 90 per cent of the absolute increase in CO_2 emissions. It is evident from these figures that the problem of global CO_2 emissions is becoming increasingly concentrated in the non-OECD area as a result of two factors — its higher growth rate in energy consumption and the fact that its fuel mix is currently more skewed towards the higher carbon-emitting fuels.

On a regional basis, the share of non-OECD CO_2 emissions is in almost direct proportion to the share of regional TPES. Hence, as Table 13 shows, by far the largest proportion of non-OECD emissions continues to originate from the USSR (31 per cent in 1989), China (21 per cent) and eastern Europe (14 per cent). Emissions from this group of countries alone were equal to about 74 per cent of emissions from the OECD region in 1989. On an individual country basis only emissions from the USA exceed those from the USSR and China. Together, these three countries account for over 50 per cent of global CO_2 emissions. When the next seven largest "emitters" are added (including India and Poland from the non-OECD), the share of total emissions rises to 71 per cent. When the first 20 countries are included, the share is 84 per cent and rises to 90 per cent when

30 countries are considered. Beyond this group of 30, no individual country accounts for more than 0.5 per cent of global emissions. These figures are significant in the context of devising a strategy to limit global emissions of CO_2 and other greenhouse gases as they indicate that the co-operation of only a relatively limited group of countries might achieve substantial gains. This would also act to limit the financial implications of any commitments by developed countries in the area of technology transfer or other forms of assistance designed to help developing countries limit their emissions of CO_2, although the costs of such a programme would still be extremely high.

Table 13: **Energy-Related CO_2 Emissions**
(mtc)

	1973 mtc	1973 %	1989 mtc	1989 %	2005[1] mtc	2005[1] %
Non-OECD	1736	40	3102	53	5330	59
USSR	685	(39)	973	(31)	1454	(27)
E. Europe	331	(19)	420	(14)	590	(11)
China	273	(16)	664	(21)	1110	(21)
Asia-Pacific	166	(10)	419	(14)	933	(18)
Latin America	154	(9)	253	(8)	437	(8)
Middle East	47	(3)	169	(5)	502	(9)
Africa	80	(5)	190	(6)	304	(6)
OECD	2609	60	2793	47	3730	41
WORLD	4345	100	5895	100	9060	100

1   Includes bunkers, petrochemical feedstocks and non-energy uses, which are excluded from the historical years

Source: IEA Secretariat

The largest contributors to CO_2 emissions in absolute terms are not necessarily the highest contributors on a per capita basis. The USSR and Poland, for example, fall somewhat in rank to eighth and tenth respectively on this scale of non-OECD countries, but China and India fall much further to 30th and 37th respectively. The countries responsible for the highest per capita emissions are generally in the Middle East, characterised by small populations and large, energy-intensive industrial sectors. Most of the countries of eastern Europe are also important in this regard. This factor will inevitably complicate any global response to the emissions problem as it raises difficult questions of equity between nations. It also indicates that emissions from some of the largest absolute sources (China, India, Indonesia) will almost certainly rise as per capita energy demand increases.

In terms of the fuel share of non-OECD emissions, it is clear that the combustion of coal dominates the emission structure, accounting for 49 per cent of total emissions in 1989. Almost 60 per cent of these emissions (and 35 per cent of world coal emissions) were generated by the USSR and China. The eastern European countries contributed a substantial 19 per cent of the total and the Asia-Pacific region a further 14 per cent. Oil combustion was responsible for 33 per cent of total CO_2 emissions in the same year and gas combustion for 17 per cent. In the case of oil, the USSR alone accounted for 30 per cent of non-OECD emissions (and 13 per cent of world emissions), with Latin America responsible for 18 per cent and the Asia-Pacific region for 17 per cent. In the natural gas area, the USSR was again the major contributor, with 63 per cent of the total. In line with developments in the structure of TPES, the share of emissions from the combustion of coal and oil has decreased over the period since 1973, while that from the combustion of gas has risen.

When considered by energy end-use sector, non-OECD CO_2 emissions have been extremely stable over time. In 1988, 53 per cent of emissions were generated by the industry sector, 16 per cent by transport and 31 per cent by "other" sectors, including agriculture, commercial and public service and residential activities. This structure differs significantly from that of the OECD in the same year, where transport accounted for 25 per cent of energy related CO_2 emissions and "other" sectors for 27 per cent. It is probably to be expected that the non-OECD structure will evolve more in line with that of the OECD over time as development processes increase the demand for motorised transport. Such a shift is likely to slow the overall increase in emissions as the oil-based transport sector is generally less "carbon intensive" than the industry sector where coal can be an important fuel source.

When forecasts of energy-related CO_2 emissions are considered, it is apparent that most of the trends of the recent period are likely to intensify. Based on the IEA's World Energy Outlook[1], the non-OECD is expected to increase its share of global CO_2 emissions to about 59 per cent by 2005, leaving the OECD countries responsible for just over 40 per cent — roughly equivalent to their forecast shares of TPES in the same year. In terms of the fuel structure of non-OECD emissions, the share of coal is likely to fall further to 43 per cent, oil to 34 per cent, with the difference being made up by emissions from gas at 23 per cent of the total. In 2005, the largest share of emissions is still expected to be generated by the USSR (27 per cent of the total), followed by China (21 per cent), Asia-Pacific (18 per cent), eastern Europe (11 per cent) and the Middle East (9 per cent).

POTENTIAL POLICY RESPONSES TO ENVIRONMENTAL ISSUES

The range of responses available to energy-related environmental problems will vary according to the nature and scale of the problem, as well as the availability of financial resources for their implementation, but will include one or more of the following:

- pollution control based on the use of add-on technologies;
- "clean" energy technologies;
- improved energy efficiency;
- fuel substitution;
- structural economic transformation; and
- lower economic growth.

In the case of OECD countries, the first two methods have been widely employed to reduce conventional local and regional pollution and, hence, acid rain. They include the use of scrubbers on power generating stations and catalytic converters on cars, and the introduction of high compression lean burn engines and fluidised bed combustion for steam raising. Significant progress has also been made in OECD countries over the last 15 years in achieving greater energy efficiency and fuel substitution, particularly away from oil.

---

1. See Chapter IV for a summary of the key assumptions on which this outlook is based.

OECD energy intensity (i.e. total energy use per unit of real GDP) fell by about 25 per cent between 1973 and 1989, while oil intensity (oil use per unit of real GDP) fell by about 40 per cent over the same period. Efficiency improvements and fuel substitution in the OECD have mostly been pursued on economic and energy security grounds and it is only more recently that they have come to be vehicles for the pursuance of environmental objectives. These two processes are seen increasingly as complementary measures to add-on and clean energy technologies and, in a number of cases, as being the only mechanisms available for effectively reducing certain types of pollutants such as greenhouse gases. Structural economic transformation in OECD economies, while not a policy response to environmental problems, has also been responsible for major gains in energy efficiency and, hence, for the alleviation of some environmental pressures. Lower overall economic growth would also have the same effect but is not considered a viable policy option.

When most non-OECD countries are considered, all of the above options for responding to environmental problems are likely to be subject to more severe constraints than would be applicable in the developed world. Add-on and clean energy technologies, for example, can involve substantial investments and potentially additional operating costs. The limited financial resources of many nations will reduce their flexibility to take maximum advantage of the available technologies and lower-cost options will probably be chosen before more effective but expensive solutions.

Fuel substitution efforts can raise specific problems in countries with a substantial energy resource base. Some of the largest non-OECD emitters of energy-related pollutants such as SO_2, NO_X and greenhouse gases are those countries, including those in eastern Europe and China and India, with large coal reserves and industrial structures and power generation systems based on coal consumption. Fuel substitution opportunities in these circumstances are less flexible as, although they might represent a declining proportion of total energy consumption, indigenous resources will continue to be used to satisfy demand. Where alternative energy supplies are available and economically viable, however, it can be expected that some substitution efforts will be pursued. A large number of developing countries, for example, contain undeveloped reserves of natural gas and it is possible that concerns about environmental problems will spur interest in their development. Foreign technology and financing are likely to be necessary, however, if these resources are to be effectively exploited. As capital and technology generally flow to where the rate of return is highest, it will be critical for non-member governments to remove many of the existing impediments to foreign investment if this fuel-substitution potential is to be realised. Potential also exists for further development of renewable energy resources, although this too will probably require extensive foreign involvement at both the research and development stage and in the transfer of technology.

It is also worth noting the particular role of nuclear power in efforts to diversify the fuel structure. Nuclear electricity, because it emits no SO_2, NO_X or greenhouse gases, provides a response to the environmental problems associated with fossil fuel consumption which has received strong support from some member countries. To date, however, nuclear has made a much smaller contribution to total energy supply in non-OECD countries and in the post-Chernobyl period, nuclear expansion plans have been scaled back quite significantly. In eastern Europe and the USSR, in particular, nuclear programmes are in a state of flux with major cancellations, interruptions of construction, temporary closures and in some instances permanent shutdowns. While safety and other related considerations are extremely important, it should be recognised that a major shift away from the nuclear power option in non-OECD countries could constrain their efforts to diversify fuel sources away from carbon-intensive fossil fuels. An appropriate response would include co-operation and assistance from OECD countries to ensure that the highest levels of safety in both plant operations and waste disposal are available to those non-member countries choosing the nuclear option.

The strategy of switching away from energy sources held to be environmentally unfriendly is likely to be complicated by its effect of making "dirtier" fuels less expensive on international markets. Conversely, increased competition for "clean" fuels such as gas and light oil is likely to result in an increase in their price. As a result, dirtier, energy-intensive types of activities may migrate increasingly to those areas of the world where environmental restrictions are less severe — this would inevitably lead to an increasing concentration of these activities in the non-OECD region.

Improved energy efficiency is likely to become an increasingly important policy objective in many non-OECD countries, particularly in eastern Europe, the USSR and China, where efficiencies are especially low. Overall, however, energy intensities are expected to remain higher than in even the least efficient OECD economies. As discussed in an earlier chapter, a range of pressures will act to actually increase intensities in many areas of the developing world. These include high population growth, increased urbanisation, greater transport demands, accelerating industrial development and improving standards of living. These factors will be more pronounced in the least developed countries, while in the more developed energy demand will grow in response to increasing demand for personal comfort. In countries with increasing per capita income, demand for energy rises beyond that necessary to satisfy essential needs and there is a corresponding increase in demand to satisfy inessential requirements. This is generally manifested in an increasing array of electrical appliances, increased space conditioning and private transport. Growing energy demand in many non-OECD countries, therefore, will stem largely from economic and social factors and cannot be solely or predominantly attributed to the absence of efficiency improvements in the production and use of energy, although this will also contribute to a higher level of energy demand. A further impediment to improved energy efficiency will be the inefficient energy pricing systems which persist in many non-OECD countries. Subsidised prices, especially for electricity, provide no incentive to conserve energy at the consumer level nor to invest in more efficient energy production and distribution technologies. They will inevitably lead to higher levels of energy consumption and, at a broader level, to a misallocation of an economy's resources.

The likelihood that structural economic transformation will contribute to the amelioration of environmental problems in the non-OECD region varies enormously across countries. For many of these countries, especially those in Africa and parts of Asia-Pacific as well as China, continued economic development is likely to lead to increasing levels of industrialisation. In other countries, there will be a continuing transition from light, labour-intensive industries to heavier, capital- and energy-intensive sectors. In both these cases, the pattern of industrial development will inevitably lead to increasing demand for energy. In only relatively few non-OECD countries, principally the newly industrialising economies of Asia with more mature industrial structures, is the economy likely to shift towards more knowledge- and technology-intensive activities with lower energy requirements. As is the case with OECD countries, lower economic growth is unlikely to be a policy response to environmental problems. Indeed, without sustained growth it is unclear how developing countries will be able to provide for environmental protection. Growth will be essential for financing new and less polluting infrastructure, funding energy research and development and adapting technologies to the needs of individual countries.

The problems created by the competing priorities of economic growth and energy consumption were in sharp evidence in the negotiations over the Montreal Protocol, where developing countries were unwilling to ignore low-cost growth opportunities which had been exploited by the industrialised world for many years. These issues are emerging again in the current process of developing a protocol on global climate change and emissions of greenhouse gases where questions relating to the absolute level and fuel structure of energy consumption will inevitably be involved.

Probably more than any other, the global warming issue highlights the increasing commonality of energy interests between OECD and non-member countries. While stratospheric ozone depletion is also a truly global issue, the problem of climate change will require an even greater degree of multi-lateral co-operation. This is because greenhouse gases result from a wider range of activities than those of CFCs and affect the interests of a broader group of producers and consumers. In addition, the CFC problem has been dealt with largely through the development of less harmful substitutes and, hence, has left end-uses mostly unchanged. Energy-related solutions to the climate change problem will demand a more complex set of responses which might alter both the fuel structure and the way in which energy is produced and consumed.

Current international policy responses to the problem, co-ordinated by the Intergovernmental Negotiating Committee on the Framework Convention for Climate Change (INC), recognise these explicitly multi-lateral interests. The Ministerial Declaration of the Second World Climate Conference, for example, which was signed by

137 countries in late 1990, provides a summary of the policy stance on climate change as it currently stands. The document suggests that the goal of international action should be to hold greenhouse gases in the atmosphere to a safe level and that achieving such a goal will require a concerted international response initiated without delay, despite scientific and other uncertainties. It also maintains that developed countries should lead the way by reducing their emissions of greenhouse gases and points out that developing countries will require financial and technological co-operation to participate meaningfully in meeting international climate objectives.

The continuing process of INC meetings is expected to resolve some of the issues raised above. In terms of meeting their own objectives, however, it would be inappropriate for IEA member countries to limit their response to global environmental problems to participation in such international fora. The extensive experience of IEA member countries and their expertise in areas such as energy efficiency and clean energy technologies can be productively used to help non-members integrate energy-related environmental considerations into their energy policies and strategies.

STATISTICAL ANNEX

Table A1: **World Total Primary Energy Supply, 1973-1989**
Mtoe

	1973	1975	1980	1985	1989
Africa	89	101	138	183	220
Asia-Pacific	195	215	300	386	509
Latin America	207	226	303	330	376
Middle East	62	73	124	190	230
China	264	314	412	516	650
Eastern Europe	349	378	458	482	480
USSR	839	929	1111	1252	1362
Non-OECD	2005	2235	2846	3340	3828
OECD	3411	3262	3622	3640	3987
World	5416	5497	6468	6980	7815

Share of World Total (%)

	1973	1975	1980	1985	1989
Africa	1.6	1.8	2.1	2.6	2.8
Asia-Pacific	3.6	3.9	4.6	5.5	6.5
Latin America	3.8	4.1	4.7	4.7	4.8
Middle East	1.1	1.3	1.9	2.7	2.9
China	4.9	5.7	6.4	7.4	8.3
Eastern Europe	6.4	6.9	7.1	6.9	6.1
USSR	15.5	16.9	17.2	17.9	17.4
Non-OECD	37.0	40.7	44.0	47.9	49.0
OECD	63.0	59.3	56.0	52.1	51.0
World	100	100	100	100	100

Note: Because of rounding, totals and sub-totals may not exactly equal the sums of their individual components

Source: IEA Secretariat

Table A2: **World Coal Supply, 1973-1989**
Mtoe

	1973	1975	1980	1985	1989
Africa	40	44	53	75	87
Asia-Pacific	75	87	108	152	189
Latin America	9	10	13	19	22
Middle East	1	1	1	3	3
China	203	234	307	405	513
Eastern Europe	220	228	259	286	270
USSR	306	320	322	297	301
Non-OECD	852	924	1063	1235	1385
OECD	715	688	835	931	972
World	1567	1612	1898	2166	2357

Share of World Total (%)

	1973	1975	1980	1985	1989
Africa	2.6	2.7	2.8	3.5	3.7
Asia-Pacific	4.8	5.4	5.7	7.0	8.0
Latin America	0.6	0.6	0.7	0.9	0.9
Middle East	0.1	0.1	0.1	0.1	0.1
China	13.0	14.5	16.2	18.7	21.8
Eastern Europe	14.0	14.1	13.6	13.2	11.5
USSR	19.5	19.8	17.0	13.7	12.8
Non-OECD	54.4	57.3	56.0	57.0	58.8
OECD	45.6	42.7	44.0	43.0	41.2
World	100	100	100	100	100

Note: Because of rounding, totals and sub-totals may not exactly equal the sums of their individual components

Source: IEA Secretariat

Table A3: **World Oil Supply, 1973-1989**
Mtoe

	1973	1975	1980	1985	1989
Africa	43	49	67	80	94
Asia-Pacific	109	114	163	178	233
Latin America	159	172	219	213	239
Middle East	43	51	93	136	149
China	53	68	89	93	114
Eastern Europe	85	96	117	99	104
USSR	322	358	438	433	433
Non-OECD	815	908	1186	1232	1367
OECD	1885	1744	1803	1580	1726
World	2700	2652	2989	2812	3093

Share of World Total (%)

	1973	1975	1980	1985	1989
Africa	1.6	1.8	2.2	2.8	3.0
Asia-Pacific	4.0	4.3	5.4	6.3	7.5
Latin America	5.9	6.5	7.3	7.6	7.7
Middle East	1.6	1.9	3.1	4.8	4.8
China	2.0	2.6	2.9	3.3	3.7
Eastern Europe	3.1	3.6	3.9	3.5	3.4
USSR	11.9	13.5	14.7	15.4	14.0
Non-OECD	30.2	34.2	39.7	43.8	44.2
OECD	69.8	65.8	60.3	56.2	55.8
World	100	100	100	100	100

Note: Because of rounding, totals and sub-totals may not exactly equal the sums of their individual components

Source: IEA Secretariat

Table A4: **World Natural Gas Supply, 1973-1989**
Mtoe

	1973	1975	1980	1985	1989
Africa	3	4	12	22	32
Asia-Pacific	6	7	15	27	47
Latin America	30	31	50	67	75
Middle East	17	21	30	51	77
China	5	7	12	11	12
Eastern Europe	40	48	69	79	80
USSR	196	228	316	461	555
Non-OECD	297	347	503	717	878
OECD	679	652	727	702	774
World	976	999	1230	1419	1652

Share of World Total (%)

	1973	1975	1980	1985	1989
Africa	0.3	0.4	1.0	1.6	1.9
Asia-Pacific	0.6	0.7	1.2	1.9	2.8
Latin America	3.1	3.1	4.1	4.7	4.5
Middle East	1.7	2.1	2.4	3.6	4.7
China	0.5	0.7	1.0	0.8	0.7
Eastern Europe	4.1	4.8	5.6	5.6	4.8
USSR	20.1	22.8	25.7	32.5	33.6
Non-OECD	30.4	34.7	40.9	50.6	53.1
OECD	69.6	65.3	59.1	49.4	46.9
World	100	100	100	100	100

Note: Because of rounding, totals and sub-totals may not exactly equal the sums of their individual components

Source: IEA Secretariat

Table A5: **World Total Primary Energy Production, 1973-1989**
Mtoe

	1973	1975	1980	1985	1989
Africa	350	312	408	425	473
Asia-Pacific	181	201	268	350	432
Latin America	319	281	385	458	500
Middle East	1098	1028	994	604	923
China	267	326	429	573	679
Eastern Europe	289	306	340	373	354
USSR	958	1086	1337	1490	1644
Non-OECD	3462	3540	4160	4273	5005
OECD	2154	2140	2491	2756	2895
World	5616	5680	6651	7029	7900

Share of World Total (%)

	1973	1975	1980	1985	1989
Africa	6.2	5.5	6.1	6.0	6.0
Asia-Pacific	3.2	3.5	4.0	5.0	5.5
Latin America	5.7	4.9	5.8	6.5	6.3
Middle East	19.6	18.1	14.9	8.6	11.7
China	4.8	5.7	6.5	8.2	8.6
Eastern Europe	5.1	5.4	5.1	5.3	4.5
USSR	17.1	19.1	20.1	21.2	20.8
Non-OECD	61.6	62.3	62.5	60.8	63.4
OECD	38.4	37.7	37.5	39.2	36.6
World	100	100	100	100	100

Note: Because of rounding, totals and sub-totals may not exactly equal the sums of their individual components

Source: IEA Secretariat

Table A6: **World Coal Production, 1973-1989**
Mtoe

	1973	1975	1980	1985	1989
Africa	40	45	72	105	106
Asia-Pacific	74	90	102	127	155
Latin America	6	7	8	14	22
Middle East	1	1	1	1	0
China	204	236	304	427	517
Eastern Europe	226	237	260	284	273
USSR	316	330	334	308	314
Non-OECD	867	946	1080	1265	1385
OECD	678	705	832	887	954
World	1545	1651	1912	2152	2339

Share of World Total (%)

	1973	1975	1980	1985	1989
Africa	2.6	2.7	3.8	4.9	4.5
Asia-Pacific	4.8	5.5	5.3	5.9	6.6
Latin America	0.4	0.4	0.4	0.7	0.9
Middle East	0.1	0.1	0.1	0	0
China	13.2	14.3	15.9	19.8	22.1
Eastern Europe	14.6	14.4	13.6	13.2	11.7
USSR	20.5	20.0	17.5	14.3	13.4
Non-OECD	56.1	57.3	56.5	58.8	59.2
OECD	43.9	42.7	43.5	41.2	40.8
World	100	100	100	100	100

Note: Because of rounding, totals and sub-totals may not exactly equal the sums of their individual components

Source: IEA Secretariat

Table A7: **World Oil Production, 1973-1989**
Mtoe

	1973	1975	1980	1985	1989
Africa	298	254	310	271	303
Asia-Pacific	92	90	119	134	152
Latin America	274	231	304	346	365
Middle East	1073	999	960	549	841
China	55	78	108	127	140
Eastern Europe	24	24	24	22	18
USSR	431	493	606	598	610
Non-OECD	2247	2170	2431	2048	2429
OECD	662	613	722	820	760
World	2909	2783	3153	2868	3189

Share of World Total (%)

	1973	1975	1980	1985	1989
Africa	10.2	9.1	9.8	9.4	9.5
Asia-Pacific	3.2	3.2	3.8	4.7	4.8
Latin America	9.4	8.3	9.6	12.1	11.4
Middle East	36.9	35.9	30.4	19.1	26.4
China	1.9	2.8	3.4	4.4	4.4
Eastern Europe	0.8	0.9	0.8	0.8	0.6
USSR	14.8	17.7	19.2	20.9	19.1
Non-OECD	77.2	78.0	77.1	71.4	76.2
OECD	22.8	22.0	22.9	28.6	23.8
World	100	100	100	100	100

Note: Because of rounding, totals and sub-totals may not exactly equal the sums of their individual components

Source: IEA Secretariat

Table A8: **World Natural Gas Production, 1973-1989**
Mtoe

	1973	1975	1980	1985	1989
Africa	8	11	21	43	57
Asia-Pacific	9	13	33	60	85
Latin America	29	31	52	67	75
Middle East	24	28	32	53	80
China	5	7	12	11	12
Eastern Europe	36	40	46	50	40
USSR	195	239	360	520	644
Non-OECD	307	369	555	803	992
OECD	683	645	681	623	666
World	990	1014	1236	1426	1658

Share of World Total (%)

	1973	1975	1980	1985	1989
Africa	0.8	1.1	1.7	3.0	3.4
Asia-Pacific	0.9	1.3	2.7	4.2	5.1
Latin America	2.9	3.1	4.2	4.7	4.5
Middle East	2.4	2.8	2.6	3.7	4.8
China	0.5	0.7	1.0	0.8	0.7
Eastern Europe	3.6	3.9	3.7	3.5	2.4
USSR	19.7	23.6	29.1	36.5	38.9
Non-OECD	31.0	36.4	44.9	56.3	59.9
OECD	69.0	63.6	55.1	43.7	40.1
World	100	100	100	100	100

Note: Because of rounding, totals and sub-totals may not exactly equal the sums of their individual components

Source: IEA Secretariat

Table A9: **World Hydroelectricity Production, 1973-1989**
Mtoe

	1973	1975	1980	1985	1989
Africa	3	3	6	5	5
Asia-Pacific	5	7	11	16	19
Latin America	10	12	20	29	38
Middle East	0	1	1	1	1
China	3	4	5	8	10
Eastern Europe	3	3	5	5	5
USSR	13	13	18	21	21
Non-OECD	37	43	65	84	98
OECD	82	91	100	111	109
World	119	134	165	195	207

Share of World Total (%)

	1973	1975	1980	1985	1989
Africa	2.5	2.2	3.6	2.6	2.4
Asia-Pacific	4.2	5.2	6.7	8.2	9.2
Latin America	8.4	9.0	12.1	14.9	17.9
Middle East	0	0	0.1	0.1	0.5
China	2.5	3.0	3.0	4.1	4.8
Eastern Europe	2.5	2.2	3.0	2.6	2.4
USSR	8.1	9.7	10.9	10.3	10.1
Non-OECD	31.1	32.0	39.4	43.1	47.3
OECD	68.9	67.9	60.6	56.9	52.7
World	100	100	100	100	100

Note: Because of rounding, totals and sub-totals may not exactly equal the sums of their individual components

Source: IEA Secretariat

Table A10: **World Nuclear Energy Production, 1973-1989**
Mtoe

	1973	1975	1980	1985	1989
Africa	0	0	0	1	3
Asia-Pacific	1	1	4	13	22
Latin America	0	1	1	2	2
Middle East	0	0	0	0	0
China	0	0	0	0	0
Eastern Europe	0	1	6	13	18
USSR	3	10	19	44	55
Non-OECD	4	13	29	73	100
OECD	49	87	157	316	406
World	53	100	186	389	506

Share of World Total (%)

	1973	1975	1980	1985	1989
Africa	0	0	0	0.3	0.6
Asia-Pacific	1.9	1.0	2.2	3.3	4.3
Latin America	0	1.0	0.5	0.5	0.4
Middle East	0	0	0	0	0
China	0	0	0	0	0
Eastern Europe	0	1.0	3.2	3.3	3.6
USSR	5.7	10.0	10.2	11.3	10.9
Non-OECD	7.5	13.0	15.6	18.8	19.8
OECD	92.5	87.0	84.4	81.2	80.2
World	100	100	100	100	100

Note: Because of rounding, totals and sub-totals may not exactly equal the sums of their individual components

Source: IEA Secretariat

GLOSSARY

ASEAN	Association of South East Asian Nations: Brunei, Indonesia, Malaysia, Philippines, Thailand, Singapore
bbl	barrel = 42 US gallons = 34.97 Imperial gallons = 158.99 litres
bd	barrels per day
CADDET	Centre for the Analysis and Dissemination of Demonstrated Energy Technologies
CFC	chlorofluorocarbons
CH_4	methane
CMEA	Council for Mutual Economic Assistance
CO	carbon monoxide
CO_2	carbon dioxide
Coal	Includes all coal, both primary (including hard coal and lignite) and derived fuels (including patent fuel, coke oven coke, gas coke, BKB, coke oven gas and blast furnace gas)
DAE	dynamic Asian economies: Hong Kong, Malaysia, Republic of Korea, Singapore, Taiwan, Thailand
EBRD	European Bank for Reconstruction and Development
EC	European Community
ESMAP	World Bank/UNDP/Bilateral Aid Energy Sector Management Assistance Program
ETDE	Energy Technology Data Exchange
GDP	gross domestic product
GW	gigawatt
GWh	gigawatt hour
IBRD	International Bank for Reconstruction and Development
IEA	International Energy Agency
IMF	International Monetary Fund

INC	International Negotiating Committee on the Framework Convention for Climate Change
LNG	liquefied natural gas
LPG	liquefied petroleum gas
mbd	million barrels per day
mt	million metric tons
mtc	million metric tons of carbon
Mtoe	million metric tons of oil equivalent
MW	megawatt
NIE	newly industrialising economies: Hong Kong, Republic of Korea, Singapore, Taiwan
NO_x	oxides of nitrogen
N_2O	nitrous oxide
O_3	ozone
OECD	Organisation for Economic Co-operation and Development
Oil	crude oil, refinery feedstocks, natural gas liquids and petroleum products
R&D	research and development
R/P	reserves production ratio
SO_2	sulphur dioxide
TPES	total primary energy supply
TWh	terawatt hour
UNDP	United Nations Development Programme
UNEP	United Nations Environment Programme
VOC	volatile organic compounds
WHO	World Health Organization
WMO	World Meteorological Organization

SELECTED BIBLIOGRAPHY

Alcamo, J., R. Shaw and L. Hordijk (eds) (1989). *The Rains Model of Acidification: Science and Strategies in Europe.* International Institute for Applied Systems Analysis. Kluwer Academic Publishers, Netherlands.

Asian Development Bank [ADB] (1991). *Environmental Considerations in Energy Development.* Asian Development Bank.

Baldwin, I. (1985). "Acid Rain: A Global Perspective". *The Environment Professional.* (7: 3).

Benedick, R.E., A. Chayes, D. Lashof, J. Mathews, W. Nitze, E. Richardson, J. Sebenius, P. Thacher and D. Wirth (1991). *Greenhouse Warming: Negotiating a Global Regime.* World Resources Institute, Washington D.C.

British Petroleum Company (1991). *BP Statistical Review of World Energy, June 1991.*

Commissariat à l'Energie Atomique [CEA] (1991). *Les Centrales Nucleaires dans le Monde, 1991.* CEA, Paris.

Economic Commission for Europe [ECE] (1991). *Economic Survey of Europe in 1990-1991.* United Nations, New York.

Goldemberg, J, T. Johansson, A. Reddy and R. Williams (1987). *Energy for Development.* World Resources Institute, Washington D.C.

Goldemberg, J, T. Johansson, A. Reddy and R. Williams (1987). *Energy for a Sustainable World.* World Resources Institute, Washington D.C.

Grubb, M.J. (1989). *The Greenhouse Effect: Negotiating Targets*. Royal Institute for International Affairs, London.

Grubb, M.J. (1990). *Energy Policies and the Greenhouse Effect. Volume One: Policy Appraisal*. Royal Institute for International Affairs, London.

Grubb, M.J., P. Brackley, M. Ledic, A. Mathur, S. Rayner, J. Russell and A. Tanabe (1991). *Energy Policies and the Greenhouse Effect. Volume Two: Country Studies and Technical Options*. Royal Institute for International Affairs, London.

Haugland, T. and K. Roland (1990). *Energy, Environment and Development in China*. Centre for Economic Analysis [ECON] and The Fridtjof Institute [FNI], Oslo.

Imran, M. and P. Barnes (1990). *Energy Demand in the Developing Countries: Prospects for the Future*. World Bank Staff Commodity Working Paper Number 23, The World Bank, Washington D.C.

International Energy Agency [IEA] (1987a). *Energy Balances of OECD Countries, 1970-1985*. OECD, Paris

International Energy Agency [IEA] (1987b). *Energy Conservation in IEA Countries*. OECD, Paris.

International Energy Agency [IEA] (1989). *World Energy Statistics and Balances, 1971-1987*. OECD, Paris

International Energy Agency [IEA] (1990). *Energy and the Environment: Policy Overview*. OECD, Paris

International Energy Agency [IEA] (1991a). *Energy Balances of OECD Countries, 1980-1989*. OECD, Paris

International Energy Agency [IEA] (1991b). *Energy Efficiency and the Environment*. OECD, Paris.

International Energy Agency [IEA] (1991c). *Energy in Non-OECD Countries*. OECD, Paris

International Energy Agency [IEA] (1991d). *Energy Policies of IEA Countries: 1990 Review.* OECD, Paris.

International Energy Agency [IEA] (1991e). *Energy Policies - Poland: 1990 Survey.* OECD, Paris.

International Energy Agency [IEA] (1991f). *Energy Statistics and Balances of Non-OECD Countries.* OECD, Paris

International Energy Agency [IEA] (1991g). *Natural Gas: Prospects and Policies.* OECD, Paris.

International Energy Agency [IEA] and the Organisation for Economic Co-operation and Development [OECD] (1991). *Greenhouse Gas Emissions: The Energy Dimension.* OECD, Paris.

International Monetary Fund [IMF] (various issues). *International Financial Statistics.* IMF, Washington D.C.

International Monetary Fund [IMF] (various issues). *World Economic Outlook.* IMF, Washington D.C.

Jones, D., (1989). "Urbanization and Energy Use in Economic Development," *The Energy Journal.* (10: 4).

Leach, G., L. Jarass, G. Obermair and L. Hoffman (1986). *Energy and Growth: A Comparison of 13 Industrial and Developing Countries.* Butterworths, England.

Leach, G. and M. Gowen (1987). *Household Energy Handbook: An Interim Guide and Reference Manual.* World Bank Technical Paper Number 67, The World Bank, Washington D.C.

Organisation for Economic Co-operation and Development [OECD] (1991a). *The State of the Environment.* OECD, Paris.

Organisation for Economic Co-operation and Development [OECD] (1991b). *Environmental Indicators: A Preliminary Set.* OECD, Paris.

Organisation for Economic Co-operation and Development [OECD] (various issues). Economic Outlook. OECD, Paris.

United Nations Environment Programme [UNEP] and World Health Organization [WHO] (1988). *Global Environment Monitoring System: Assessment of Urban Air Quality.* UNEP/WHO, London.

U.S. Agency for International Development [USAID] (1990). *Greenhouse Gas Emissions and the Developing Countries: Strategic Options and the U.S.A.I.D. Response — A Report to Congress.* US Agency for International Development, Washington D.C.

World Bank (1989). *Sub-Saharan Africa: From Crisis to Sustainable Growth.* The World Bank, Washington D.C.

World Bank (1990). *World Debt Tables 1990-91: External Debt of Developing Countries.* The World Bank, Washington D.C.

World Bank (1991a). *Annual Report, 1991.* The World Bank, Washington D.C.

World Bank (1991b). *Trends in Developing Economies, 1991.* The World Bank, Washington D.C.

World Bank (1991c). *World Tables.* Johns Hopkins University Press, Baltimore.

World Bank (various issues). *World Development Report.* Oxford University Press, New York.

World Resources Institute [WRI], in collaboration with The United Nations Environment Programme [UNEP] and The United Nations Development Programme [UNDP] (1990). *World Resources 1990-91: A Guide to the Global Environment.* Oxford University Press, New York.

WHERE TO OBTAIN OECD PUBLICATIONS – OÙ OBTENIR LES PUBLICATIONS DE L'OCDE

Argentina – Argentine
Carlos Hirsch S.R.L.
Galeria Güemes, Florida 165, 4° Piso
1333 Buenos Aires Tel. 30.7122, 331.1787 y 331.2391
Telegram: Hirsch-Baires
Telex: 21112 UAPE-AR. Ref. s/2901
Telefax:(1)331-1787

Australia – Australie
D.A. Book (Aust.) Pty. Ltd.
648 Whitehorse Road, P.O.B 163
Mitcham, Victoria 3132 Tel. (03)873.4411
Telefax: (03)873.5679

Austria – Autriche
OECD Publications and Information Centre
Schedestrasse 7
D-W 5300 Bonn 1 (Germany) Tel. (49.228)21.60.45
Telefax: (49.228)26.11.04
Gerold & Co.
Graben 31
Wien I Tel. (0222)533.50.14

Belgium – Belgique
Jean De Lannoy
Avenue du Roi 202
B-1060 Bruxelles Tel. (02)538.51.69/538.08.41
Telex: 63220 Telefax: (02) 538.08.41

Canada
Renouf Publishing Company Ltd.
1294 Algoma Road
Ottawa, ON K1B 3W8 Tel. (613)741.4333
Telex: 053-4783 Telefax: (613)741.5439
Stores:
61 Sparks Street
Ottawa, ON K1P 5R1 Tel. (613)238.8985
211 Yonge Street
Toronto, ON M5B 1M4 Tel. (416)363.3171
Federal Publications
165 University Avenue
Toronto, ON M5H 3B8 Tel. (416)581.1552
Telefax: (416)581.1743
Les Publications Fédérales
1185 rue de l'Université
Montréal, PQ H3B 3A7 Tel.(514)954-1633
Les Éditions La Liberté Inc.
3020 Chemin Sainte-Foy
Sainte-Foy, PQ G1X 3V6 Tel. (418)658.3763
Telefax: (418)658.3763

Denmark – Danemark
Munksgaard Export and Subscription Service
35, Nørre Søgade, P.O. Box 2148
DK-1016 København K Tel. (45 33)12.85.70
Telex: 19431 MUNKS DK Telefax: (45 33)12.93.87

Finland – Finlande
Akateeminen Kirjakauppa
Keskuskatu 1, P.O. Box 128
00100 Helsinki Tel. (358 0)12141
Telex: 125080 Telefax: (358 0)121.4441

France
OECD/OCDE
Mail Orders/Commandes par correspondance:
2, rue André-Pascal
75775 Paris Cédex 16 Tel. (33-1)45.24.82.00
Bookshop/Librairie:
33, rue Octave-Feuillet
75016 Paris Tel. (33-1)45.24.81.67
  (33-1)45.24.81.81
Telex: 620 160 OCDE
Telefax: (33-1)45.24.85.00 (33-1)45.24.81.76
Librairie de l'Université
12a, rue Nazareth
13100 Aix-en-Provence Tel. 42.26.18.08
Telefax: 42.26.63.26

Germany – Allemagne
OECD Publications and Information Centre
Schedestrasse 7
D-W 5300 Bonn 1 Tel. (0228)21.60.45
Telefax: (0228)26.11.04

Greece – Grèce
Librairie Kauffmann
28 rue du Stade
105 64 Athens Tel. 322.21.60
Telex: 218187 LIKA Gr

Hong Kong
Swindon Book Co. Ltd.
13 - 15 Lock Road
Kowloon, Hong Kong Tel. 366.80.31
Telex: 50 441 SWIN HX Telefax: 739.49.75

Iceland – Islande
Mál Mog Menning
Laugavegi 18, Pósthólf 392
121 Reykjavik Tel. 15199/24240

India – Inde
Oxford Book and Stationery Co.
Scindia House
New Delhi 110001 Tel. 331.5896/5308
Telex: 31 61990 AM IN
Telefax: (11)332.5993
17 Park Street
Calcutta 700016 Tel. 240832

Indonesia – Indonésie
Pdii-Lipi
P.O. Box 269/JKSMG/88
Jakarta 12790 Tel. 583467
Telex: 62 875

Ireland – Irlande
TDC Publishers – Library Suppliers
12 North Frederick Street
Dublin 1 Tel. 744835/749677
Telex: 33530 TDCP EI Telefax: 748416

Italy – Italie
Libreria Commissionaria Sansoni
Via Benedetto Fortini, 120/10
Casella Post. 552
50125 Firenze Tel. (055)64.54.15
Telex: 570466 Telefax: (055)64.12.57
Via Bartolini 29
20155 Milano Tel. 36.50.83
La diffusione delle pubblicazioni OCSE viene assicurata
dalle principali librerie ed anche da:
Editrice e Libreria Herder
Piazza Montecitorio 120
00186 Roma Tel. 679.46.28
Telex: NATEL I 621427
Libreria Hoepli
Via Hoepli 5
20121 Milano Tel. 86.54.46
Telex: 31.33.95 Telefax: (02)805.28.86
Libreria Scientifica
Dott. Lucio de Biasio 'Aeiou'
Via Meravigli 16
20123 Milano Tel. 805.68.98
Telefax: 800175

Japan – Japon
OECD Publications and Information Centre
Landic Akasaka Building
2-3-4 Akasaka, Minato-ku
Tokyo 107 Tel. (81.3)3586.2016
Telefax: (81.3)3584.7929

Korea – Corée
Kyobo Book Centre Co. Ltd.
P.O. Box 1658, Kwang Hwa Moon
Seoul Tel. (REP)730.78.91

Malaysia/Singapore – Malaisie/Singapour
Co-operative Bookshop Ltd.
University of Malaya
P.O. Box 1127, Jalan Pantai Baru
59700 Kuala Lumpur
Malaysia Tel. 756.5000/756.5425
Telefax: 757.3661
Information Publications Pte. Ltd.
Pei-Fu Industrial Building
24 New Industrial Road No. 02-06
Singapore 1953 Tel. 283.1786/283.1798
Telefax: 284.8875

Netherlands – Pays-Bas
SDU Uitgeverij
Christoffel Plantijnstraat 2
Postbus 20014
2500 EA's-Gravenhage Tel. (070 3)78.99.11
Voor bestellingen: Tel. (070 3)78.98.80
Telex: 32486 stdru Telefax: (070 3)47.63.51

New Zealand – Nouvelle-Zélande
GP Publications Ltd.
Customer Services
33 The Esplanade - P.O. Box 38-900
Petone, Wellington
Tel. (04)685-555 Telefax: (04)685-333

Norway – Norvège
Narvesen Info Center - NIC
Bertrand Narvesens vei 2
P.O. Box 6125 Etterstad
0602 Oslo 6 Tel. (02)57.33.00
Telex: 79668 NIC N Telefax: (02)68.19.01

Pakistan
Mirza Book Agency
65 Shahrah Quaid-E-Azam
Lahore 3 Tel. 66839
Telex: 44886 UBL PK. Attn: MIRZA BK

Portugal
Livraria Portugal
Rua do Carmo 70-74, Apart. 2681
1117 Lisboa Codex Tel.: 347.49.82/3/4/5
Telefax: (01) 347.02.64

Singapore/Malaysia – Singapour/Malaisie
See "Malaysia/Singapore" – Voir «Malaisie/Singapour»

Spain – Espagne
Mundi-Prensa Libros S.A.
Castelló 37, Apartado 1223
Madrid 28001 Tel. (91) 431.33.99
Telex: 49370 MPLI Telefax: 575.39.98
Libreria Internacional AEDOS
Consejo de Ciento 391
08009 - Barcelona Tel. (93) 301-86-15
  Telefax: (93) 317-01-41
Llibreria de la Generalitat
Palau Moja, Rambla dels Estudis, 118
08002 - Barcelona Telefax: (93) 412.18.54
Tel. (93) 318.80.12 (Subscripcions)
(93) 302.67.23 (Publicacions)

Sri Lanka
Centre for Policy Research
c/o Mercantile Credit Ltd.
55, Janadhipathi Mawatha
Colombo 1 Tel. 438471-9, 440346
Telex: 21138 VAVALEX CE Telefax: 94.1.448900

Sweden – Suède
Fritzes Fackboksföretaget
Box 16356, Regeringsgatan 12
103 27 Stockholm Tel. (08)23.89.00
Telex: 12387 Telefax: (08)20.50.21
Subscription Agency/Abonnements:
Wennergren-Williams AB
Nordenflychtsvägen 74, Box 30004
104 25 Stockholm Tel. (08)13.67.00
Telex: 19937 Telefax: (08)618.62.32

Switzerland – Suisse
OECD Publications and Information Centre
Schedestrasse 7
D-W 5300 Bonn 1 (Germany) Tel. (49.228)21.60.45
Telefax: (49.228)26.11.04
Librairie Payot
6 rue Grenus
1211 Genève 11 Tel. (022)731.89.50
Telex: 28356
Subscription Agency – Service des Abonnements
Naville S.A.
7, rue Lévrier
1201 Genève Tél.: (022) 732.24.00
Telefax: (022) 738.48.03
Maditec S.A.
Chemin des Palettes 4
1020 Renens/Lausanne Tel. (021)635.08.65
Telefax: (021)635.07.80
United Nations Bookshop/Librairie des Nations-Unies
Palais des Nations
1211 Genève 10 Tel. (022)734.14.73
Telex: 412962 Telefax: (022)740.09.31

Taiwan – Formose
Good Faith Worldwide Int'l. Co. Ltd.
9th Floor, No. 118, Sec. 2
Chung Hsiao E. Road
Taipei Tel. 391.7396/391.7397
Telefax: (02) 394.9176

Thailand – Thaïlande
Suksit Siam Co. Ltd.
1715 Rama IV Road, Samyan
Bangkok 5 Tel. 251.1630

Turkey – Turquie
Kültur Yayinlari Is-Türk Ltd. Sti.
Atatürk Bulvari No. 191/Kat. 21
Kavaklidere/Ankara Tel. 25.07.60
Dolmabahce Cad. No. 29
Besiktas/Istanbul Tel. 160.71.88
Telex: 43482B

United Kingdom – Royaume-Uni
HMSO
Gen. enquiries Tel. (071) 873 0011
Postal orders only:
P.O. Box 276, London SW8 5DT
Personal Callers HMSO Bookshop
49 High Holborn, London WC1V 6HB
Telex: 297138 Telefax: 071 873 2000
Branches at: Belfast, Birmingham, Bristol, Edinburgh,
Manchester

United States – États-Unis
OECD Publications and Information Centre
2001 L Street N.W., Suite 700
Washington, D.C. 20036-4910 Tel. (202)785.6323
Telefax: (202)785.0350

Venezuela
Libreria del Este
Avda F. Miranda 52, Aptdo. 60337, Edificio Galipán
Caracas 106 Tel. 951.1705/951.2307/951.1297
Telegram: Libreste Caracas

Yugoslavia – Yougoslavie
Jugoslovenska Knjiga
Knez Mihajlova 2, P.O. Box 36
Beograd Tel.: (011)621.992
Telex: 12466 jk bgd Telefax: (011)625.970

Orders and inquiries from countries where Distributors
have not yet been appointed should be sent to: OECD
Publications Service, 2 rue André-Pascal, 75775 Paris
Cedex 16, France.

Les commandes provenant de pays où l'OCDE n'a pas
encore désigné de distributeur devraient être adressées à :
OCDE, Service des Publications, 2, rue André-Pascal,
75775 Paris Cédex 16, France.

75880-7/92

OECD PUBLICATIONS – 2, rue André-Pascal, 75775 PARIS CEDEX 16
PRINTED IN FRANCE
(61 92 01 1) ISBN 92-64-13618-5 – N° 45893 1992